ADVANCE PRAISE FOR *THE STORY OF TECHNOLOGY*

"The clearest, most comprehensive overview of technology, and the government's role in developing it, that I've seen. Rich with insights, it's a must-read for entrepreneurs looking for traction with the government, and for anyone seeking to understand what technology's all about."

General Wesley K. Clark
former NATO Supreme Allied Commander Europe

"In describing technology as being both a constant companion throughout human existence, and, at times, an unwanted interloper, Gerstein provides interesting bookends to his analysis. He argues that the future of technology is not preordained and that the type of future that humankind desires must be carefully considered as part of an important policy debate."

Secretary Thomas Ridge
first secretary of the Department of Homeland Security and
former governor of Pennsylvania

THE STORY
OF TECHNOLOGY

HOW WE GOT HERE AND WHAT THE FUTURE HOLDS

DANIEL M. GERSTEIN

Prometheus Books
59 John Glenn Drive
Amherst, New York 14228

Cover image © Shutterstock
Cover design by Nicole Sommer-Lecht
Cover design © Prometheus Books

Inquiries should be addressed to
Prometheus Books
4501 Forbes Boulevard, Suite 200
Lanham, Maryland 20706
www.rowman.com

Distributed by NATIONAL BOOK NETWORK

Library of Congress Cataloging-in-Publication Data

Names: Gerstein, Daniel M., 1958- author.
Title: The story of technology : how we got here and what the future holds / by Daniel M. Gerstein.
Description: Amherst, New York : Prometheus Books, 2019. | Includes bibliographical references and index.
Identifiers: LCCN 2019011629 (print) | LCCN 2019013706 (ebook) |
    ISBN 9781633885790 (ebook) | ISBN 9781633885783 (hardcover)
Subjects: LCSH: Technology—Social aspects—History. | Technological innovations—History. Classification: LCC T14.5 (ebook) | LCC T14.5 .G457 2019 (print) | DDC 609—dc23
LC record available at https://lccn.loc.gov/2019011629

It is difficult to make predictions, especially about the future.
—Attributed to Niels Bohr and Yogi Berra
(among others)

CONTENTS

APPENDICES

INTRODUCTION

Throughout the development of this book—from the initial stages to the final edits—technology was a constant interloper. Everywhere I went, someone was talking about a new technology or a problem that had been solved through scientific discovery. Such immersion made for a wealth of topics to be covered but also made it hard to know when to put the pen down.

I was in a wine bar, and a rather large group was having a spirited discussion about the use of blockchain, debating about its merits and how it could improve cybersecurity. The next day, a friend wanted to talk to me about technologies for smart city applications. A few days later, I received an email from a colleague asking what I knew about Bitcoin—in particular, the pros and cons of investing in the speculative online currency. When I came home that evening, the nightly news reported about a hacker who had gained access to data from soldiers' personal Fitbits. The report described how the hacker had written code that would allow him to hack the network so that the movements of military service members around the world could be tracked. A news report immediately after the Fitbit story lamented the efficacy of the annual influenza vaccine, highlighting that the 2018 strain match was not particularly good.

I did not seek out these discussions on technology. Rather, they found me. These were not technologists but ordinary people who had been exposed to various technologies that were either a part of their daily lives or that they were thinking about or trying to understand. In some cases, they were trying to determine if the technologies should become part of their lives. The discussions highlighted that each was approaching these technologies with a mix of excitement and trepidation, not fully understanding the science or technology behind many of these areas, yet at the same time wanting to know more.

It was at this point I decided the story of technology—the past, present, and future—was one that should be told. Through this telling, perhaps technology could be made less opaque and placed in its proper context. For throughout the history of humankind, technology has been a constant companion, propelling us forward and at times showing our dark side. We are also at an interesting point in the development of technology, at which the convergence of multiple fields has the potential to fundamentally alter the human existence.

My vantage point for the study of technology comes from the perspective of an operator. I have spent most of my professional life considering national security and defense issues. My training, education, and experiences have also largely been concentrated on such applications. Whether in uniform, in industry, or as a senior civilian in the US government, my reflections on the use of technology have largely been from an operational perspective.

Interspersed with these operational experiences have been educational opportunities to consider the evolution of technology, how it affects our society and the world in which we live, and what might be the possibilities for the future.

I have come to believe that technology has been a central organizing construct throughout the history of humankind. We see this in our everyday tools and conveniences as well as in the manner and speed at which we consume and exchange information. However, over the past two hundred years, the rate of technological development has accelerated exponentially. The confluence of technologies led by information technology, communications, and transportation has been the driving force behind a series of monumental changes that we have labeled globalization.

We are witnessing a symbiotic relationship between technologies where advances in one area contribute to growth in other endeavors. As technologies develop, they lead to other technologies and even new fields. Just over the past fifty years, we have seen space travel to the moon and the formation of a new domain called cyber, and we are possibly on the verge of redefining what it means to be human through a mix of biotechnology, artificial intelligence, and the Internet of Things advancements.

The overwhelming pace of technological development has made the world seem smaller, increased competition between nations, and even greatly changed the dynamics of human interactions. We are now more intertwined across global societies and know more about each other and the physical world in which we live.

Ominously, the rate of change of technology, either individual technologies or in combinations, shows no evidence of slowing down. In fact, such change and technological development appears to be accelerating. For me, the analogy has been the Disney World ride in which one sits in a car that moves at a fixed pace while projections along the walls create the perception that one is speeding along, moving first by horse and carriage, then in an automobile, then in a rocket ship zooming through the galaxy. Just as the Disney ride moves at a fixed pace, humans have biological processes that have remained relatively constant—the beating of a heart, the speed at which we move—yet our surroundings have accelerated, placing greater demands on us in areas such as cognition and crisis response.

We are living in a nonlinear world where combinations of technologies have become far greater than the sum of their parts. In this way, we signify that the function is no longer additive but, rather, multiplicative—or perhaps even exponential. We are witnessing new generations of technology in rapid succession across many technology areas. Consider how rapidly new generations of cellular telephones are delivered to market or how autonomous systems are becoming a reality, encroaching into nearly every aspect of our lives. It will likely not be too long until we become a driverless society.

It is against this backdrop that we will consider technology development. Looking at the past provides a perspective on how far we have progressed. Understanding where we are today allows us to take stock of the progress we have made. Looking toward the future allows us to think of the possibilities.

In this future, understanding which technologies will be the most important, how to combine them to gain maximum benefit, and technology's overall effect on society will be important to the future of humankind. Just as in millennia past, when humans used early technology to build tools to survive, future generations must also build the tools that will ensure their survival.

This book examines the future of technology and proposes a methodology to better understand how technology is likely to evolve in the future. It will not attempt to pick individual technology winners and losers but, rather, will seek to provide a structured way to think about future technologies. Looking through this lens, we will identify the attributes that a future successful technology will seek to emulate and identify pitfalls that a technology developer should seek to avoid. This process will also help us think about the impact of technology on individuals and societies.

Future technological development has the potential to alter the traditional relationships between nations and societies, and societies and individuals. Undoubtedly in this developmental process, there will be those who argue for going slowly, while others are likely to call for an even faster pace. How this delicate balance proceeds will be important to the types of societies that emerge.

Inevitably, questions about the potential for managing technology will arise. Will the mechanisms that have been developed and are used today remain relevant (or successful) in this new age of technology? Specifically, will laws, policies, regulations, standards, and codes of ethics developed in a different era remain useful to managing technology in the future? Or will new mechanisms need to be developed?

In this discussion, some may seek to label certain technologies as being not worthy of examination, identify them as significant dual-use threats with both civilian and military applications, or paint them as inherently dangerous. I reject this characterization of technology and would submit that it is not the technology but how it is used by individuals, organizations, and nations that determines whether it becomes a threat—or an opportunity. Furthermore, in considering technology's role in the future, one only need look at history to understand that halting the development of a technology is irrational, self-defeating, and, anyway, not possible.

Humankind has always sought to advance, and technology has been one of the facilitators of and accelerators for this progress. Technology for survival created opportunities to gain better understandings of the natural world, and thus for societies to grow. These understandings helped build the world of today.

This book was motivated by several personal experiences. First, my observations about the importance of technology in the security and defense field have been foundational. The history of warfare contains numerous examples of specialized technologies designed to improve the efficiency of armies in battle. It also contains examples of technologies developed for civilian or peaceful purposes being applied in warfare with great effect. Perhaps not as well understood is that many of the technologies originally developed for use in warfare have later been adapted for nonmilitary uses with great benefit to society.

Second, my experiences serving in government in the Department of Homeland Security in the Science and Technology Directorate caused me to directly confront questions surrounding how technologies are developed and what leads to successful transitions that fulfill a need. I also saw firsthand that successful technology development has as much to do with getting people to use it as with the technology itself. It may be terrific, but if the operator cannot or will not use it, it is irrelevant.

Finally, on many occasions, friends and colleagues have asked questions about the development of technology. Some are simply looking to know more about the field. Others are asking for practical purposes as they look to integrate technological solutions into a business deal. Still others are looking to better understand how a novel, dual-use, or disruptive technology is likely to shape an outcome.

This book has been developed with these experiences in mind. The book consists of three parts: "Foundations," "Managing Technology," and "The Future." In "Foundations," the basics of technology are considered. What is the difference between science and technology, research and development, innovation and transformation? Do these distinctions matter any longer, given how in our society today, they are largely used interchangeably? In "Managing Technology," the current systems of control are considered to understand how effective they have been and how to think about technologies that will be developed in the future. The final part of the book, "The Future," examines the roles and responsibilities of those who use technology (which is all of us) in relation to those who develop technology.

In reading this book, try to put yourself in the perspective of the main character in the story. After all, we all have vast and varied experiences with technology, both positive and negative. Try to envision how your life would be different without many of the tools and gadgets we have at our disposal. And also try to place yourself in the future and think of the possibilities that advances in technology are likely to bring. Ask yourself whether technology could actually be controlled— and if so, how, to what end, and with what downsides?

And so, our journey begins . . .

PART I

FOUNDATIONS

Technology has advanced more in the last thirty years than in the previous two thousand. The exponential increase in advancement will only continue.

—Niels Bohr

Modern technology has become a total phenomenon for civilization, the defining force of a new social order in which efficiency is no longer an option but a necessity imposed on all human activity.

—Jacques Ellul

CHAPTER 1

IN THE BEGINNING . . .

As we go about our daily lives, it has become nearly impossible to remain divorced from the technologically enabled world of the twenty-first century. Seemingly everywhere we are confronted with some kind of new technological or scientific marvel. Short of divesting of our digital footprint and turning our clocks back several hundred years, our prospects for halting or reversing the march of technology are dim.

At the same time, technology has become both an enabler for society and a potential threat from those who would endeavor to misuse it. The proliferation—some call it the democratization—of technology has made it far more widely available. At times, it has placed statelike capabilities into the hands of individuals. Consider cyber, which has a low barrier for entry and allows the user access to critical infrastructure systems, personal data, and the ability to project information across the globe in milliseconds. It is no overstatement to say that any technology that has been developed has at some point in history been misused. Such is the nature of technology, and of humankind.

For much of the history of humankind, development has proceeded along a linear path. The evolution of humans largely proceeded on a steadily increasing trajectory. A complementary increase in technologies aided in this development, supported the progress of the species, and led to the building of societies. With very real purpose, humankind developed a capacity to feed itself. The development of spoken and written languages allowed for communicating with each other and recording human history. Our quest for discovery resulted in increased trade and travel and intermixing of peoples and cultures. The introduction of mathematics by the Babylonians and Egyptians in around 3,000 BCE provided a basis for logic and problem-solving.

For the past two thousand years, humankind has seemingly been on a technological tear. Major cities have been formed, transportation systems have been developed that allow masses of people to routinely traverse the globe in a matter of hours, and communications networks connect us instantaneously and beam information around the globe at the speed of light. Today, humankind stands on the verge of being able to manipulate the very essence of life.

Just over the course of the last two hundred years, the pace of discovery and the resultant changes in societies has been dramatic. The Agricultural Revolution ended, and the Industrial Revolution is now in the rearview mirror. In the twentieth century alone, the automobile and winged aircraft were introduced and became common modes of transportation. Humans have landed on the moon and now live above us on the international space station. The last quarter century saw the invention of a new domain, cyberspace.

We are in the Information Age, and the rate of change continues to increase across a wide swath of technology areas. Terms to describe technologies that did not exist a handful of years ago are now commonplace: cyber, artificial intelligence (AI), Internet of Things (IoT), augmented reality and virtual reality, nanotechnology, quantum computing, neural networks, and 3-D printing are just some of the new areas we are encountering.

The early years of the twenty-first century continue to demonstrate the dominance of technology in this new era. Advances in a broad range of technologies have had fundamental impacts on our societies, both positive and negative. Social media and twenty-four-hour news cycles have allowed us to know more about our world than ever before. At the same time, they have been credited with electing a president in the United States and threatening societies overseas. Genetically modified organisms have become the enemy in some parts of the world—for example, in the European Union. Cyberattacks threaten to make the internet unusable and have exposed the private records of more than 150 million Americans to foreign entities and criminal elements.

AT THE INTERSECTION OF HUMANKIND AND TECHNOLOGY

Today, we are rarely separated from technology as we go about our daily lives. Regardless of where a person goes on the planet, it is likely that she will have the advantages of technology. The clothes she is wearing have likely been designed to be water resistant or optimized for the climatic conditions. The cellular telephone she carries provides position location information, a means of communication, and an ability to catalog her travels, to name just a few of the functions. If she is going camping, the backpack allows for conveniently and efficiently carrying cumbersome loads. And the pocket knife she carries, which consists of multiple tools contained within a convenient package, allows her to survive in the wilderness and perform functions from cutting wood to eating with utensils. But ask yourself, "What is technology, and where does it come from?"

At its core, technology is about solving problems. It is the practical application of knowledge and has been around for as long as humans have been in existence. Early humans as far back as 1.76 million years ago developed hand axes and large cutting tools to assist them in their daily lives.[1] Weapons were used for hunting, making the task of acquiring food less dangerous and more efficient. After the hunt, other tools were used in breaking down the kill so that it could be consumed. Utensils were used for cooking the meat. Pottery was created for the storage of water, grains, and berries. These early tools allowed humans to grow and prosper. They also served as the foundations for the technology of today. In fact, the evolution of humankind and technology are intertwined.

As humankind evolved from subsistence hunting and gathering to modern societies, technology supported this process. The early subsistence gathering eventually evolved into the development of agriculture as a means of acquiring food. Evidence of humans consuming wild grain can be traced back more than a hundred thousand years, but the deliberate and planned cultivation of grains and domestication came far later, around fifteen thousand years ago.[2] The earliest pottery has been discovered in China dating back twenty thousand years. It was likely used for storage as well as cooking and steaming, including baking pita-like foods.[3] The crockery also allowed for a more mobile society, as grains and water could be more readily transported.

Many refer to this period as the Agrarian Age or the beginning of the Agricultural Revolution. Regardless of terminology, the march toward the development of modern societies took a major leap forward as a result and led to the development of implements that made agriculture—both growing crops and domesticating livestock—more efficient. In turn, this provided greater yields to feed hungry peoples, grow stronger humans, and build societies.

The nature of humankind has been to seek better ways to do things, to increase the efficiency and effectiveness of a task. In agriculture, that has translated to development of new techniques to grow, harvest, and process grains; promote the domestication, growth, and health of livestock; and find more efficient ways to feed populations. Irrigation made for greater availability of arable lands for growing crops. Farming tools were developed to more efficiently plow and plant fields. Wetlands were drained for growing crops. Crop rotation began in the sixteenth century.[4] Modern farming techniques including the introduction of machinery coincided with the beginnings of the Industrial Age.

Each of the mechanical enhancements was designed to solve a real-world problem. The cotton gin was designed to rapidly separate fibers from seeds. The tractor was introduced to more rapidly plow the fields and plant and harvest the crops. Long before people understood genetics, plant breeding was introduced that resulted in the development of cross crops with traits and genes that were beneficial for crop yield or survival in harsh environments. Some were designed to be resistant to drought or to pests. The results of the introduction of technology to the field of agriculture are clear. By one account, in the 1940s, a single farmer could feed nineteen people. By the year 2000, one farmer could feed 155 people.[5]

In discussing progress in agriculture, one account cited five technologies as being essential for this increase in productivity. The plow, which was introduced in approximately 3500 BCE in Egypt, allowed for more rapidly tilling the soil. By 1855, John Deere Company was selling thirteen thousand plows per year. The introduction of the tractor, which was aided by the development of the steam engine in the 1870s, allowed for more rapid tilling of the soil. The efficiency this brought also reduced the cost of producing crops. The combine, which could harvest quickly, made what was fundamentally a labor-intensive process highly mechanized and less costly. The introduction of fertilizer in the 1890s protected crops against disease and led to an increase in yields. Finally, biotechnology, including the selective breeding of animals and plants with highly desirable traits, has led to increased yields and efficiencies. As of 2012, "94 percent of cotton, 93 percent of soybeans, [and] 88 percent of corn in the United States were planted using biotech seeds."[6]

Increases in technology have resulted in related changes in the labor markets. With the introduction of technology, there has been a decrease in the number of people required in agriculture, with corresponding increases in the industry and service sectors. During this same period, crop yields have continued to increase. For example, in 1840, approximately 70 percent of the US workforce was engaged in agriculture. The other 30 percent were working in industry or in the services sector. By 2012, the US workforce in agriculture was down to 1.4 percent. Industry, which included mining, construction, and manufacturing, rose steadily from 1840–1955 and peaked at approximately 38 percent of the workforce. Interestingly, the number of people in the industrial sector has continued to decline and now stands at slightly less than 20 percent of the US labor market. The reduction is largely due to robotics and automation.

So far, we have been talking about technology, or the practical application of knowledge. But it is also interesting to consider when science—

the quest for fundamental understanding of natural phenomena—began. Identifying the point in human history when science came into being is both challenging and controversial. Evidence suggests that at a very early point in our development, humans became interested in gaining a greater understanding of the world in which we live.

Some of the earliest science can be traced to the field of astronomy. Locations have been identified in Africa and Europe that were used to track the moon's phases as early as 35,000 BCE. More recently, at Crathes Castle in Scotland, twelve pits were discovered, dating from ten thousand years ago, that track the phases of the moon and lunar months. Many consider this formation a step toward the formal construction of time. Whether these measurements were used to tell time in a practical sense or were seeking to gain a fundamental understanding of a natural phenomenon is not clear. In the first case, it would reflect a technological development, while in the second case, an attempt to gain scientific understanding.[7]

Throughout the history of humankind, an interesting phenomenon can be seen. Observations regarding cause and effect have at times led to the technology being applied long before the science is understood. Such was the case in the development of a vaccine for the smallpox virus in the 1790s. While Edward Jenner did not understand the science behind a smallpox virus infection, he did observe that milkmaids never became ill with smallpox, but instead were harmlessly infected with cowpox. He inferred that this was due to their close work with cows and hypothesized that a cowpox infection could offer immunity against the deadlier smallpox virus. Acting on this observation, he inserted pus from a local milkmaid infected with cowpox into the arm of a local boy, James Phillips. Several days later, Jenner tested his theory by exposing the boy to live smallpox virus. As surmised, Phillips did not come down with smallpox. Interestingly, when Jenner attempted to publish his findings, he was turned down, as it was deemed that further proof was necessary before the article would be accepted.[8] Even the science of Jenner's day required more than anecdotal evidence to verify his claim.

In the history of humankind, technological enhancements have often served as inflection points after which societies became fundamentally changed. For example, the use of the first tools was an inflection point that fundamentally altered the prospects for the species' survival and continued evolution. Other examples of inflection points related to technology include the development of language and mathematics, the printing press, mass production, the nuclear era, and the internet. Some may argue that religion could fall into this category as well. While I would not object to such a characterization, it is beyond the scope of this book.

Just as no technology is either all good or all bad, neither are the resulting inflection points. One could easily go through this previous list of technologies and identify where an inflection point has resulted in both good and not-so-good outcomes. Nuclear energy and atomic weapons come to mind in this regard. Of course, some might be tempted to argue that even nuclear weapons and the strategic stability they brought were beneficial in preventing a major global war between the East and West during the Cold War.

KEY THEMES

The book considers technology's effect on society in the past, present, and future. Along the way, several key themes will be developed.

As the capabilities of humankind have increased, the quest for gaining a fundamental understanding of the world has driven greater scientific study, which in turn has fueled greater technological achievements. Through this discovery, we are learning more about the universe and each other in leaps and bounds.

Our world is more connected than ever—some might even say smaller—and events in one part of the world have implications for nations and peoples thousands of miles away. Advances in information, communications, and cyber technologies are allowing for sharing of ideas, data, and information in real time. Entire movements, as evidenced by 2010's Arab Spring, have gone from a spark to a raging inferno in a matter of hours and days.

Dual-use technologies are allowing individuals to have statelike capabilities. Areas of science and technology that once were the exclusive domain of nations, such as Global Positioning Systems (GPS), have become parts of our daily lives. The growth of multinational corporations complicates placing controls on or managing science and technology proliferation.

Laws and treaties, other international organizations such as the United Nations and World Health Organization, coalitions of the willing, and unilateral actions—both individually and in combination—have historically been used to shape outcomes, influence national and international actions, and enhance national security. With the proliferation of science and technology across the globe, the effectiveness of many of these institutions is being called into question.

In many areas, advances are so rapid they are outpacing policy makers' ability to place adequate controls on them. In the absence of policies and regulations, technologies are continuing to move forward. Decision makers in government and industry are frequently confronted with how to respond to a new advancement before its implications have even been fully assessed. Some get it right, and others don't.

SEIZING THE OPPORTUNITIES AND PREPARING FOR THE CHALLENGES

As scientific discoveries and technological advances continue at this breakneck pace, having the foresight to properly and rapidly identify the challenges and opportunities that await will be imperative.

The technology superhighway is littered with those who either have been passed at light speed by newer technology or failed to read the signs and adapt. The prize will go to those who are able to look at the future and anticipate the likely changes in technology. In this journey, there are no recipes for success. Rather, there are indications of when changes are occurring within a technology area or when a novel application of technology might signal a new beginning—or an impending end. Being able to detect these changes in enough time to affect outcomes must become a priority for future technologists.

Each new technology has evolved in its own way. However, commonalities in evolution exist that can be useful to observe and understand. Technologies today do not stand alone but, rather, are combinations of other technologies that have been brought together for practical purposes. Numerous examples of these combinations exist. Cars consist of multiple technologies combined into systems and further integrated into a compact conveyance that allows drivers to operate safely in moving from place to place. Mobile computing makes use of computer science, radio electronics, electricity, transistors, software development, antenna technology, and batteries, to name a few. Advances in biotechnology are based on the convergence of several technologies including information technology (IT), without which such breakthroughs as high-throughput sequencing and bioinformatics would not have been possible.

This book also examines how others have sought to assess the future of technology. What have been the successes and failures? What are the elements that will likely contribute to understanding how a technology will mature? After considering what lessons have been learned in assessing technologies, a methodology will be proposed for examining individual technologies and their resultant impact on particular areas of activity.

Finally, in many respects, the idea of managing technology essentially comes down to a question of balance. What is the proper balance between allowing scientific discovery and technological advancement to go largely unfettered versus imposing constraints that would stifle their progress? What are the benefits versus the risks of developing the technology? What is the proper balance between security and privacy? Does more need to be done to protect US research and development and intellectual property, or should the benefits be shared globally?

In reflecting on these questions, consider the following scenarios. Do they cause you concern, or are they just part of the progress of human advancement and technology development? How you answer speaks volumes about your individual preferences in areas such as privacy and security.

- the use of Internet of Things technologies to constantly monitor traffic, the maintenance on your car, and your location
- the widespread substitution of autonomous systems for humans in the workplace
- disease research using gain-of-function methods to increase disease virulence to aid in the development of treatments and cures
- targeted sales advertisements based on your internet search history
- the use of biometric capabilities for screening at the speed of life, which would virtually eliminate waiting in lines but leave vast trails of digital footprints
- allowing your DNA to be used for population-wide studies for disease prevention that would ultimately lead to truly personalized medicine but could also be used for decisions affecting education, employment, and insurance
- the use of gene editing to make changes to the human genome (1) for curing disease, (2) for enhancing performance, or (3) for altering fundamental human qualities
- the development of lethal artificial intelligence and autonomous systems capabilities for use on the battlefield
- the theft of more than $1 trillion in intellectual property from US military aircraft programs

One caveat is in order. This book is not intended to be an exhaustive study of technology throughout the history of time. Nor is it intended to cover all technologies and their contribution to human development in chronological order. Furthermore, it does not attempt to look at individual technologies, except as such an examination contributes to the overall understanding of science and technology's impact on individuals and societies.

Our goal throughout the book will be to take a long view of technology, examining the past and the present in the hopes of gaining insights into the future. Doing so allows us to understand the possibilities, as well as the challenges likely to be encountered. While the past and present have been and are currently being written, the future is ours to shape. The role of society and the technologist—using this term in its broadest sense to include scientists, technology developers, and individuals who rely on technology in their everyday lives—will be to shape a future that is hospitable to humankind.

CHAPTER 2
PERSPECTIVES AND DEFINITIONS

Think back to challenges you have faced. They don't have to be big challenges—maybe just getting along in everyday life. Perhaps you are cooking and need to cut up the ingredients for your recipe. Maybe you want to get a message to your friends and family, and you reach for your cell phone. Or perhaps you need to find directions to a restaurant and quickly get on the computer to download them.

Technology—or perhaps, more specifically, combinations of technologies—is now available to assist you in these daily tasks. Without even thinking about it, we reach for a technology to solve our problem. Likely, you do not even think about the technological marvels that have become the fixtures of everyday life. These technologies, evolved over millennia of human existence, have now become routine parts of our daily lives.

In this chapter, we will place technology into a historical context to demonstrate the codevelopment of humankind and technology; four technologies will serve as examples of these dependencies. We will also consider technology from the users' perspectives. My own experiences as an "accidental technologist" will be offered as a demonstration for how technology has aided and, at times, encroached upon our lives.

And finally, we will develop working definitions for science and technology; research and development; and innovation and transformation. These are the definitions that will be used throughout the remainder of the book.

HISTORICAL PERSPECTIVES

In the first chapter, the relationship between humankind and the development of technology was described as synergistic with humankind's quest for practical purposes and solutions to problems. The results were technologies that led to increased human development. In fact, technology has been around for as long as humans have. But describing the relationship in this way does not seem to go far enough. For the codevelopment of humankind and technology has been an important constant throughout recorded history.

If we examine the ages of human history, they correspond to the technologies of the day. The Stone Age (50000–3000 BCE) saw the use of wood and stone for daily tasks, weapons, and building materials. The Bronze Age (3000–1300 BCE) and Iron Age (1200–250 BCE) saw the introduction of these metals in the technologies of the day. Development of ancient societies in Egypt, India, Greece, and Rome (8000 BCE to 476 CE) introduced the concepts of societies and laws, use of technology for warfare, agriculture, building cities, and increased trade and travel. The Islamic Golden Age (750–1300 CE) led to discoveries in science and mathematics. The Renaissance (1350–1650 CE) saw a rebirth of "culture, arts, science and learning." Humankind's codevelopment of societies and technology continues into the modern era's Agrarian Age, Industrial Revolution, Nuclear Age, and Information Age.[1] In fact, throughout history we have measured human development and technological maturity using the same scale.

From the earliest days, humans have sought ways and means to survive and become self-sufficient. In the same way that humans have evolved over millions of years with our DNA recording this journey, so too has technology. The early building blocks of technology can be seen in even the most sophisticated of modern technologies.

While an exhaustive recounting of human and technology development will not be attempted, placing this codevelopment in context will provide a basis for understanding how their intertwined history is likely to affect future technology development. In this examination, the individual technologies developed over time are far less important than how they have been combined into technology systems and have led to inflection points in the history of humankind. Several examples will be considered in demonstrating how technology development can affect the trajectory of humankind: the printing press, electricity, nuclear weapons, and the internet. Each of these technologies resulted in an inflection point in human history.

The printing press provided a means to accurately share knowledge. Prior to its development, information was passed through oral recounting or handwritten documents. Neither provided the speed or accuracy to proliferate knowledge effectively or efficiently. Early printing was done using a technique called block printing that entailed carving blocks of wood, coating them with ink, and pressing sheets of paper on them to transfer the imprint.[2]

This process was far less time-consuming and more accurate than either handwriting or speech. The technique was employed by Chinese monks more than six hundred years before the first printing press was developed. In between, in the eleventh century, a method for using movable characters made of clay was developed that represented a significant improvement in speed over block printing. Advancements continued in the materials used in printing, going from clay to wood, to bronze, and finally to steel.

In 1440, a German craftsman, Johannes Gutenberg, developed the first printing press. Gutenberg experimented with various materials, configurations, and processes before developing a mechanized way to transfer the ink from movable type to paper and move the printed paper off the ink press, creating an efficient assembly line for rapid printing. Gutenberg's printing press essentially made knowledge more affordable and available. Previously, only those who could afford to pay to have a manuscript copied had access to books and, by extension, the knowledge they contained.

The printing press also promoted the rapid sharing of knowledge and contributed to a widely literate public. The effect of the printing press was perhaps disproportionately seen in the sciences, where the accurate sharing of information is crucial. In fact, the printing press has been cited as leading to the Scientific Revolution of the Enlightenment in the 1600s.[3]

The natural phenomenon of electricity was observed long before it was understood, could be characterized, or was able to be controlled. Early observations of its effects on humans from electric fish and lightning strikes provided ample evidence of electricity. The early efforts of humans to understand electricity included observations of static electricity from friction caused by rubbing amber (as early as 600 BCE), and of the relationship between electricity and magnetism (in the 1600s).

In the seventeenth and eighteenth centuries, theories and experimentation continued, with the definitive link between the theory of electricity and lightning (1752); the theory of bioelectromagnetics, which established electricity as the mechanism by which neurons passed signals to the muscles (1791); the recognition of electromagnetism (1820); and the invention of the electric motor (1821). Discovery and invention continued throughout the latter half of the eighteenth century and into the nineteenth with such developments as the electric battery, the telephone, radiotelegraphy, light bulbs, and rotating magnetic fields.[4]

It is not by happenstance that the growing use of electricity led to a second Industrial Revolution (which saw electricity begin to replace the steam power that was the hallmark of the first Industrial Revolution). Today, electricity is embedded throughout modern society and essential in all communications and infrastructure. One only need look at the impact of Superstorm Sandy in 2012 to understand how important the electrical sector is to the American way of life. The storm affected twenty-four states, including most of those along the Eastern Seaboard, and caused more than 130 deaths, massive flooding, and more than $65 billion in damages. The storm also left 8.5 million people without power, including more than a million who lacked power even a week after the storm. It required the efforts of more than seventy thousand linemen, technicians, and other workers from around the nation to help with power restoration. The storm also reinforced that the electrical sector was first among equals for critical infrastructure. Without electricity, gas stations could not operate, the financial sector could not function, restaurants could not reopen, and normalcy would not return to the lives of those affected.

The devastating loss of electrical power during successive hurricanes that made landfall in Puerto Rico in October 2017 served to further reinforce the degree to which modern societies have come to rely on electrical power. Failure to rapidly establish power exacerbated the human tragedy, as even five months after the storm, one-third of the island or nine hundred thousand people still lacked power. The delay in restoring power prevented reopening of hospitals and resulted in numerous deaths.

The development of nuclear capabilities has clearly affected the trajectory of humankind, for both military necessity and peaceful civilian use. Nuclear weapons represent an important inflection point in global security affairs, while the development of nuclear energy provided a new source of efficient, clean energy.

Nuclear power plants generate some 14 percent of the world's energy; however, some countries such as France get more than 75 percent of their energy from nuclear energy. Many other countries are seeking to incorporate nuclear power into their energy portfolios but have been limited by initial start-up costs and requirements for safeguards for nuclear material.[5] Some 450 power generation reactors and 225 research reactors are in use around the globe. Many of these research reactors serve as sources for medical isotopes used in medical applications. Another 160 or so are planned to be built for power generation.[6]

The history of the development of nuclear technology will be discussed in greater detail later in the book. However, additional discussion of the strategic implications of nuclear weapons will be discussed here, as these weapons ushered in the Nuclear Age and the associated deterrence framework that has been the "cornerstone of American security for more than seventy years."[7] During the Cold War, the United States and our allies relied on strategic, intermediate, and tactical nuclear weapons to offset the large advantages in conventional military forces held by the Soviet Union and the rest of the Warsaw Pact. It is not hyperbole to state that nuclear weapons were a foundational capability on which both sides came to rely. Around the globe today, many nations have come to see the acquisition of a nuclear weapons arsenal as a strategic imperative to assuring their security: the five declared nuclear weapons states (United States, Russia, China, Britain, and France) as well as Israel, India, Pakistan, North Korea, and Iran to name a few. Still other nonnuclear nations desire to be under a nuclear umbrella as a guarantee of their security—such is the case for the NATO alliance that benefits from this extended deterrence posture.

Today, a smaller strategic nuclear arsenal provides deterrence—and, should deterrence fail, the necessary strategic and warfighting capabilities to protect US interests in the complex twenty-first-century security environment. The US deterrence strategy seeks to "prevent attacks by influencing an adversary's decision-making, ensuring they see no gain in attacking the United States or our allies."[8] The nuclear and conventional capabilities have been crafted into a complex deterrence architecture in which assets are routinely moved around the globe to signal intentions, provide assurances to allies, and defend US interests.

The internet, like electricity, has ushered in a new era: in this case, the Information Age. For our purposes, *internet* will be used as a broad umbrella term that includes the development of computers and the early networks for passing messages as well as the global network that exists today. While the history of the development of the internet will also be presented in detail later in the book, here our goal is to highlight the synergistic relationship between the technology and the development of society and humankind.

Today, the internet has become a new domain, serving (similarly to electricity) as the lifeblood of a large majority of the global population and nearly all US citizens. We have come to rely on this technology for information about our friends, family, and society.

The internet has also become a de facto utility allowing for monitoring, assessing, and controlling critical infrastructure, communications, and financial transactions, among other uses.

Yet at times, these two uses of the internet—for information sharing and as a utility—seem a bit in conflict and perhaps even portend two alternative futures. Is it an information-sharing platform with capabilities for controlling the imperatives of daily life? Or is it an industrial control system that can also provide the capacity to share vast amounts of information across global society?

Despite the differences in how each of these four technologies—the printing press, electricity, nuclear weapons, and the internet—have evolved and the role they play in global society today, there is little doubt that these technologies have been part of or caused an inflection point in the course of human history.

OPERATIONAL PERSPECTIVES

At times, I have felt like an "accidental technologist," but I have concluded that many of us likely feel that way. The pervasiveness of technology means we are always drawing on some capability that has evolved over the course of human development or endeavoring to use a modern technology or looking for the next "bleeding edge" technology.

Throughout my professional career, I have had the opportunity to rely on, consider the applications of, and develop technologies. Some of these experiences have been smooth and successful, while others have been challenging and resulted in less than successful outcomes. In some ways, technology has been the bane of my existence, as often I have been called upon to implement technology solutions.

Through these experiences, I have come to see technology as being much broader than the application of science or knowledge, and also far more expansive than a tool, gadget, or piece of equipment. In fact, the derivation of the word *technology* in Greek includes "art, skill, cunning of hand" and implies the application of devices, methods, and practices for practical purposes. I have found this more expansive understanding of technology to be more relevant in considering technology's role in our lives. Furthermore, my perspectives on technology are most directly related to operational experiences in security and defense. Examining technology through this lens, one sees both its potential and its pitfalls.

Technology can improve the quality of our daily lives and increase the effectiveness and efficiency of the work that we do. But if misused, technology could become a dangerous capability for those with malicious intent. One only need look at the terrorist attacks of September 11, 2001, to see that the same planes that allow us to connect with people across the globe can also be used as missiles to destroy buildings and kill innocent people. Both possibilities must be considered when examining technology. And it is in this spirit that I offer the following anecdotes and some encapsulating thoughts about my view of technology.

The technologies of today are really systems of systems.

As a young lieutenant in the US Army, my job entailed installing message centers and communications systems in command centers supporting major warfighting headquarters and senior commanders. The purpose was to deliver message traffic essential for decision-making in a potential conflict. It was a highly technical field and required understanding of a broad range of subordinate technologies. My platoon had vehicles that required maintenance, power generation systems for the communications equipment, and radios and teletypes for passing message traffic. The equipment was mobile and had to be reinstalled with each move of the supported element's operations centers. Furthermore, these subsystems needed to be synchronized to operate as a component of the larger system, which ultimately included its tactical emplacement on a potential battlefield. Interestingly, years later, this unit consisting of tens of millions of dollars' worth of equipment and seventy personnel has largely been replaced by a wireless smartphone outfitted to send and receive classified messages.

Technologies can and must be combined and recombined to achieve operational purposes.

As a commander of a six-hundred-person signal battalion in the 1990s, my unit was deployed in Bosnia-Herzegovina as part of the implementation force peace operation. Over time, the goal was to replace the current tactical communications equipment with commercial communication systems provided by one of the major long-haul communications providers. The commercial system would include telephone communications and internet services. The melding of technologies—the tactical system that had been in place for six months and the new commercial system that was to be installed—required combining technologies in ways that had not necessarily ever been envisioned, and was not without some challenges. However, while these two systems were never intended to operate in the same network, they had been developed using similar communications protocols and standards. Eventually, through the efforts and technical skill of the soldiers, the two systems were combined relatively seamlessly into a single communication system.

A technology is more than the sum of a system's hardware and software components.

While serving in Operation Desert Storm with the Third Armored Division (one of the lead elements in the invasion of Iraq), my duties required the planning and coordination of a network that would eventually stretch more than 220 kilometers across three countries and be established in a one-hundred-hour period. The network was designed to provide communications capabilities for frontline units and senior commanders directing the war. One of the interesting experiences in the preparation for the attack occurred one day when four very large cases were delivered to the command post. I was asked by the senior leader on site what I was going to do with these new systems. After some inspection, I reported to the general that I had no idea what to do with the new "technology" that had been delivered. In fact, it was only after the war had ended that I would come to find out the new systems were called maneuver control systems or MCS. They were designed to provide the commander with increased situational awareness on the battlefield. This experience reinforced that technology is only useful if accompanied by the knowledge to use it. Having four large boxes that had to be carried throughout the course of the attack and for which no instruction had been provided proved to be nothing but frustrating.

Technology includes the application of science, knowledge, and methods for achieving practical purposes.

As a commander in charge of a signal brigade and immediately in the aftermath of 9/11, my unit was required to have the capability to deploy complex communication systems throughout the world rapidly and in very austere conditions. As these capabilities had not been fully fielded throughout the army, the task became to develop our own deployable communications packages. Each system required integration of a satellite terminal, telephone switching, classified and unclassified internet, and a video teleconference suite. These capabilities and technologies all existed, but not necessarily together, and certainly not in the configurations that would be needed to fulfill the missions that were envisioned. Therefore, the requirement was to develop the system architecture, procure the individual components, configure the systems, test the systems, train the soldiers, and certify the equipment as operational. The task required the combination of multiple systems of systems. For example, delivering the internet capability required combining individual technologies such as servers, routers, hubs, smart switches, and computers into a single integrated system. Each individual component needed to operate properly, and all components had to be integrated to form the overarching structure—anything less meant that the system would not function.

Technology development cannot be rushed and proceeds at its own pace. Even leadership attention and generous resourcing may not be decisive in pushing technologies to maturity.

When the Chief of Staff of the Army announced that the army would transform to become a more rapidly deployable Information Age force, I had the opportunity to serve as the director of army transformation. The goal was to bring innovative technologies and ways of operating into the force; provide the force with increased deployability, enhanced lethality, greater protection, and vastly increased situational awareness; and reduce the logistics footprint that supports the warfighting forces. The technology centerpiece of the new formation would be the future combat system (FCS). This new platform would be designed to allow the force to "see first, understand first, act first, and finish decisively." The army chief of staff wanted to begin fielding the FCS within a ten-year period while bringing forward technologies that were still in the tech base. Some were at very low technology readiness levels and would need to be significantly matured before being able to be incorporated into the new systems. Others were further along but would need to be integrated before being fielded to operational units. The plan called for spiral development of technology, whereby the system would continue to be modernized as the subordinate technologies matured. Over time, it became clear that the new technology insertions would not be ready to be fielded along these aggressive time lines and that the cost per individual system would be considerably higher than envisioned. Eventually, the FCS program would be canceled, as it was determined to be technologically infeasible and unaffordable.

This same concept was attempted by the navy in the fielding of a new aircraft carrier. The *George W. Bush* was designed to field novel technologies that were under development but had not yet been proven. The thought was to fund and begin building the aircraft carrier while maturing some of the early-stage technologies. One of these systems to be rapidly matured and inserted when the technology was ready was the electric catapult system for launching aircraft, which was to replace the steam-powered system. This plan introduced significant risk into the procurement of this new aircraft carrier. In fact, many of the technical details of this new catapult technology have still not been overcome, and the *George W. Bush* has failed to meet cost, schedule, and performance metrics.

Getting a technology to market requires more than interesting concepts—it must be able to be produced in operationally relevant numbers at a reasonable cost, and it must fulfill an operational need.

After leaving the army, I went to an industry focused primarily on security and defense, where one of my duties was to assist my corporate leadership with mergers and acquisitions by assessing technologies with an eye toward buying either the technologies or the companies. Invariably, I would run across very interesting technologies for which the developers had failed to consider how to get the technologies to market or how to mass-produce their products. In some cases, they even lacked the intellectual property rights that would allow them to monetize their technologies. This is mentioned not to disparage these companies but rather to highlight that different skill sets are required for conceiving of, developing, and integrating technologies that will eventually find their way to market.

Delivering technology means understanding one's core business model and where to go for the technology solutions.

As the Deputy Under Secretary for the Science and Technology Directorate (S&T) within the Department of Homeland Security (DHS), I was part of the leadership team with responsibility for providing research and development for the department. For example, for law enforcement technologies, the Homeland Security Act of 2002, which established the department and S&T Directorate, required the organization "to serve as the national focal point for work on law enforcement technology;

and to carry out programs that, through the provision of equipment, training, and technical assistance, improve the safety and effectiveness of law enforcement technology and improve access to such technology by Federal, State, and local law enforcement agencies." However, nowhere in this mission statement did it say that the directorate was required to actually do all the S&T development.[9] The same was true for protecting critical infrastructure, nuclear smuggling, and chemical and biological defense. In fact, most of the directorate's tasks involved establishing priorities, planning, monitoring R&D, and advising the secretary and the operational components. Even where R&D activities were being conducted, they were largely being done by external performers.

The S&T Directorate had more than 1,100 employees and a budget of more than $1 billion per year. It also consisted of laboratories, partnerships, and programs for the purpose of supporting DHS in integrating innovative technologies and providing technical support. Over time, it became clear that the directorate had been somewhat misnamed. There really wasn't much science or technology development being done within the organization. Rather, most of the employees were serving as technologists. As such, their role was to link operational requirements with those conducting the research and development. The difference is subtle but very important, as the skill sets required of the workforce and the way they interfaced with those needing help with operational problems varied greatly between scientists and technologists.

Building partnerships is essential as conducting basic and applied research, and early-stage development is largely unaffordable without relying on the existing body of knowledge and collaborating with others.

While a $1 billion-per-year budget seems like a great deal of funding, it does not go far when spread across the third-largest cabinet-level agency in the United States. Conducting basic and applied research is a costly proposition and certainly not affordable across all the missions DHS has. Therefore, determining what others had already accomplished was essential. It was also necessary to gain an understanding of what types of organizations are best at the various research and development areas.

The S&T Directorate technologists were charged with looking across their respective fields to understand what others were doing and how those efforts could be brought into the department to improve operational outcomes. It required outreach to external technology organizations, laboratories, universities, and industries, more as strategic scouts than technology developers. More frequently than not, these organizations were involved in applied research and technology development. In only a very few cases was basic research or science being conducted by partner organizations. In fact, most were looking for technologies that were either being developed or had been developed that could aid in solving a specific operational problem.

For example, the National Science Foundation, the Department of Energy laboratories, the National Institutes of Health, laboratories within the military departments of the Department of Defense, and universities all do basic and applied research. The same is true for small companies that are operating in niche areas looking to develop technologies.

In S&T, we functioned more like a large integrator looking for technologies that were in later stages of development and could be readily adapted for DHS purposes. Large integrators today rely on others for most of their basic and applied research (and science). Many of the major companies have people with technology skills actively looking for cutting-edge technologies that can be licensed and integrated into their products and offerings.

My operational experiences with technology are not necessarily unique. While my early experiences were focused on operationally employing technology, my latter experiences became directed toward developing technology solutions for others.

At the heart of all these experiences with technology has been the purposeful application of capabilities (including scientific knowledge or the collection of techniques, methods, or processes) for practical purposes.

DEFINITIONS

We have been casually mentioning several paired terms—science and technology; research and development; and innovation and transformation—without clearly defining them or highlighting their differences. In this section, we will correct this shortcoming by establishing working definitions for each.

Frequently when discussing technology with colleagues, I often find myself wondering how they are using the word. When someone asks whether a technology will work, are they talking about whether a tool will be useful in a particular application, or whether the technology is mature and able to be used, or whether a doctrinal or organizational change might be in order to solve the problem they are encountering? Depending on the answer, the person could be referring to science, technology, research, development, innovation, or transformation.

So far, technology has been loosely identified with the act of solving practical problems using devices, methods, and practices, while science has been identified as the quest for fundamental understanding of natural phenomena.

On one hand, this was appropriate, as there was little discernible distinction between what early humans were doing to solve problems versus to gain a fundamental understanding of a natural phenomenon. However, going forward, it will be useful to clearly delineate the differences between these terms. Doing so will enable our understanding of science and technology, fostering an appreciation for the relationship between them and the differences that distinguish them.

One may even hear these terms being used in combination. Frequently, one might hear someone professing the need for technological innovation. What is one looking for in this case? Is this person looking for a practical application or something that is novel, or both?

Some might attempt to delineate boundaries separating science and technology; research and development; and innovation and transformation. These exercises are not particularly fruitful, as the terms are inextricably overlapping, and a linear progression between these pairs does not exist. For example, science does not necessarily progress into technology, nor does it always lead to research and development. Innovation can occur as part of developing scientific techniques for use in technology development or by an operator using a new technology. Transforming organizations often incorporates innovation to reform processes or through the infusion of new technologies.

To an operator seeking to have a real-world (or operational) problem addressed, the distinctions likely do not seem important. Even for those involved in these discoveries and the investigation of technical areas, absolute lines can rarely be drawn. Basic and applied research could easily drift into initial stages of development. Technology development could easily encounter a problem, or unknown, that requires returning to basic and applied research.

It is through this blurry lens that includes science and technology, research and development, innovation and transformation that we will introduce clarity by defining these terms for our purposes going forward. A more detailed discussion of the historical derivation of these terms has been included in appendix A.

SCIENCE AND TECHNOLOGY

Separating these two terms over the course of human history is difficult, and even identifying with any confidence when science began is challenging. The introduction of the written word and mathematics were certainly key to development of the scientific method. Without them, making observations, taking measurements, and conducting repeatable experiments would not have been possible. So while today, the meaning of the term *scientific method* varies across fields of study, it encompasses some basic elements: (1) formulating hypotheses, (2) collecting data, (3) testing hypotheses, and (4) developing conclusions.

Our shorthand definition up to this point has been that science is the quest for fundamental understanding of natural phenomena. The derivation of the word provides some clarity but doesn't help to distinguish it from the concept of technology. Science has a long etymological history, coming from the Greek word meaning "true sense"; from the Latin *scientia*, which means "knowledge"; and from Old French meaning "knowledge, learning, application."[10]

The Science Council—an organization that provides the quality assurance program for those working in science—spent over a year looking at the definition of science and concluded, "Science is the pursuit of knowledge and understanding of the natural and social world following a systematic methodology based on evidence."[11]

As for our purposes going forward, our definition of science draws on the yearlong Science Council effort but will introduce the rigor associated with scientific discovery. Our resulting definition of science will be:

> Science is the pursuit of knowledge and understanding of the natural and social world following a systematic (or scientific) methodology based on evidence derived from observations, experimentation, analysis, and repetition of results.

Regardless of the precise definition, basically science is grounded in the use of the scientific method, which provides a means of understanding a phenomenon, organizing the body of thought surrounding how it functions, and drawing conclusions and principles regarding an area of study.

Now turning to technology, we see clearly how the terms *science* and *technology* have become intertwined and, at times, interchangeable in modern vocabulary. Even the definition of science above contains terms such as *knowledge* that earlier in our working explanations of technology we had used in describing this term. In trying to introduce a degree of separation, think of the number of times you have been watching the news and seen reporting on a new discovery, such as development of a new insulin pump. Ask yourself: is that science, or is it technology? Is there a difference, and does it really matter which one resulted in this medical advancement? More than likely, the combination of scientific discovery and technological development has resulted in many new capabilities, including the insulin pump. Understanding the need for insulin—the natural hormone produced in the pancreas that regulates glucose in the blood—relies on science, while the pump that moves the fluid and regulates the concentration of insulin provided is a mechanical technology.

Clues to the precise meaning of the word *technology* can be found in its derivation, from a Greek word that is often translated as "craftsmanship," "craft," or "art," implying that it goes far beyond simple toolmaking to include the methods used in developing and using a technology.

For our purposes going forward, the concept of technology will be applied in the broader sense, capturing the roots of the term that come from its derivation, bringing in the idea that technology is broader than applying science and knowledge and includes the concept of developing capabilities. Therefore, our definition will be:

> Technology is the application of capabilities for practical purposes. It can include inputs from scientific knowledge or the collection of techniques, methods, or processes applied in a purposeful way. The term applies across a broad range of human endeavors.

Given the meanings of science and technology, it is not surprising that the terms have frequently been used interchangeably and often as a pair.

RESEARCH AND DEVELOPMENT

Research and development, normally referred to simply as R&D, is a relatively new concept beginning in the early part of the twentieth century. Even for those working in the field, the paired terms *research* and *development* can at times appear to be interchangeable with the concepts of science and technology. *Research* seemingly aligns very nicely with *science*, while *development* and *technology* appear to have a similar comfortable alignment. However, such a simple interpretation of the terms would fail to recognize important distinctions between them.

One distinction between these two pairs of terms relates to the intended purpose of the actions. Science and technology have been described, respectively, as activities undertaken for gaining a fundamental understanding of natural phenomena and using scientific knowledge for practical applications. In contrast, for basic research conducted as part of R&D, there is normally some overarching concept of use—for example, a natural phenomenon must be understood to allow for proceeding to applied research, where that phenomenon can begin to be applied (and later exploited in the development phase). It is difficult to envision that corporations today would engage in basic research with no expectation of a return on their investment. Later in our discussions of scientific discovery, we will highlight the essential role of government in promoting basic research.

Research and development are clearly related to the purposeful activity designed to result in attainment of new capabilities for an operational or economic purpose. R&D also relates closely to activities associated with an industrial process, including production, manufacturing, and procurement (or acquisition). In fact, the term *R&D* derives from the Industrial Revolution, when companies engaged in research intending to build on scientific principles and ultimately apply their findings to developmental activities that would result in products that could be bought and sold; in other words, their focus was on moving products through a research-development-acquisition life cycle.

To demonstrate the symbiotic relationship between science and technology and research and development, the definitions of technology readiness levels provide an important starting point. Technology readiness levels (TRLs) provide a systemic approach to identifying the maturity of various technologies.[12] In this framework, the nine levels are defined as:

1. Basic principles observed and reported
2. Technology concept and/or application formulated

3. Analytical and experimental critical function and/or characteristic proof of concept

4. Component and/or breadboard validation in laboratory environment

5. Component and/or breadboard validation in relevant environment

6. System/subsystem model or prototype demonstration in a relevant environment

7. System prototype demonstration in an operational environment

8. Actual system completed and qualified through test and demonstration

9. Actual system proven through successful mission operations

The TRLs combine R&D into a single framework, where TRLs 1–3 are considered research and related closely to scientific discovery. Development entails a broad range of activities designed to mature the technology prior to fielding and actual use in an operational environment. Some might consider TRLs 4–9 to be development, while others may choose to define only 4–7 this way. In the latter case, TRLs 8–9 may be considered acquisition and thus part of the broader life cycle of technology, no longer strictly development.

It is important to note that no universal definitions of research and development exist. For example, each cabinet-level department and agency of the US government has a different definition tailored to meet its specific R&D process.

No attempt has been made to provide an exact cross-walk of the pairs S&T and R&D. As the above discussion indicates, close relationships exist between science and research and between technology and development. In the case of science and research, principles are attempting to be developed leading to a better understanding of a natural phenomenon. For technology and development, they seek to develop practical uses from undertaking these activities.

INNOVATION AND TRANSFORMATION

Innovation and transformation (I&T) are terms that frequently get interwoven with science and technology and/or research and development. However, *innovation* has a very different connotation, entailing the application of new ways of thinking and novel ideas. Innovation also involves risk-taking. It can be evolutionary, with continuous change and adaptation, or revolutionary, which results in discontinuous change that is disruptive and new. In describing innovation, one expert states, "Innovative ideas can be big or small, but breakthrough or disruptive innovation is something that either creates a new category, or changes an existing one dramatically, and obsoletes the existing market leader."[13]

Another definition states that innovation is "the process of translating an idea or invention into a good or service that creates value or for which customers will pay." Of course, this definition comes from a source specializing in business applications.[14]

One of my former bosses used to describe innovation by pointing out that there was nothing new about wheels, and there was nothing new about suitcases—but adding wheels to the suitcase revolutionized air travel.

In contrast to S&T and R&D, which are methodical processes, innovation relates more to the generation of ideas and development of combinations of technologies and capabilities in a far less structured approach. In describing innovation, innovators often talk of developing ideas for solving problems or improving efficiency. Innovations often result from whittling down the list of a wide variety of ideas to a small subset. In this way, perhaps only one or two original ideas would eventually be developed into something (i.e., an application, good, or service).

Innovation is not just about products, either. It is possible to innovate processes and gain greater efficiencies through the implementation of innovative solutions. Such process reform is really at the heart of Lean Six Sigma, which was first introduced by Toyota as a way to streamline its processes and eliminate waste. Automobile manufacturers and other product manufacturers are natural candidates for use of such techniques. When the US Army required repair of tracks damaged during combat in Iraq, the contractor implemented a Lean Six Sigma methodology to eliminate unnecessary steps in the process, reducing the time for average repairs by an order of magnitude. The contractor determined that much of the wasted processing time resulted from moving the trucks around the factory from station to station. By eliminating all but the necessary moves (such as moving the truck to and from the paint shop) and instead having the specialized teams move between the trucks, weeks were eliminated from the overall processing time.

By now, it should be clear that no single authoritative definition exists for innovation. However, certain common themes are evident. Innovation is a process for generating, promoting, and operationalizing ideas. It results from remaining connected to the operational problem that needs to be solved. Innovation cannot be allowed to become synonymous with only procurement or acquisition of new hardware or software as it is far subtler. Promoting a culture that will allow ideas to be generated, most of which are likely to be discarded, will undoubtedly be among the most important attributes for innovation to succeed.[15] At the core, innovation is about taking risks and having the freedom to pursue alternative solutions. Innovation can result in evolutionary or revolutionary changes to approaching operational requirements and developing new capabilities.

The other half of this third pair, *transformation*, is a term that describes a comprehensive change within an organization. The process of transformation can occur in any organization, industry, or government. However, the term owes a significant portion of its popularity to its use (some might say overuse) by the US military.

The word *transformation* was initially used in military circles to describe a comprehensive change that affects nearly all aspects of the organization. This is not to say that other government and civilian organizations were not undertaking meaningful change previously, but rather that the military was instrumental in coining this usage of the term.

There are certainly examples of businesses that have undertaken transformations to improve their organizational structure, processes, and outputs. And these changes envisioned through these organizational redesign processes would be no less significant than those envisioned by the military beginning in the late 1970s, when technologies associated with precision weapons and stealth and greater emphasis on rapidly deployable expeditionary forces took center stage in changes to the force.[16] In the 1990s, the military used the concept of a Revolution in Military Affairs (RMA) to rewrite doctrine, change organizational structures, and procure weapons systems. In the late 1990s, RMA was displaced by transformation.

And in fact, all the envisioned transformation initiatives by the Department of Defense and in industry would have required some amount of S&T, R&D, or innovation.[17]

Seeking to increase effectiveness and efficiency has led government, industry, and even individual consumers to pursue innovation and transformation as a top priority. In some respects, this pursuit feels more like an effort to short-circuit the S&T and R&D processes, providing access to innovative and transformative capabilities without the heavy investments required of early S&T and R&D. Whether such a strategy will be effective in the long run or serve as a substitute for traditional S&T and R&D remains an unanswered question. It is more likely that some combination of S&T, R&D, and I&T will be required to satisfy the needs of government, industry, and consumers for useful capabilities to solve real-world problems.

The importance of risk-taking in innovation and transformation cannot be overstated. On the approach to innovation, many have highlighted the path as "failing fast and failing often." The point is to develop novel ideas, give them a try, and assess rapidly whether they have utility. Innovators would not be content to spend years perfecting a single idea only to discover it had no utility for solving the operational problem. The same can be said of transformation, where change can be discomforting and even disruptive. In such cases, the transformation can be seen as threatening to the status quo. This implies the likelihood of failure can be high given organizational inertia, which can overwhelm the forces of change.

Defining the six terms with some degree of rigor is a way to better understand the similarities and differences between them. However, at this point in our journey, we can use the shorthand of technology development unless referring to a specific process directly. The frequency with which these terms are used interchangeably or in combination (e.g., scientific research, innovative technology, or transformative technology) in our daily lives argues strongly for such a shorthand.

However, by way of a disclaimer, while the distinctions between these activities might not matter as greatly to users, some experts in a field may need to retain these distinctions—for example, in science, where scientists have a very specific method for conducting their work and the rigor associated with conducting repeatable research requires a more demarcated approach. Such distinctions are also required in government-funded activities, as Congress appropriates money in specific "pots of money" such as R&D, acquisition, and operations and maintenance (or life cycle support). Moving money around in these categories requires congressional approval and therefore is tracked very closely.

SO WHERE IS ALL THE TECHNOLOGY DEVELOPMENT GETTING DONE?

It has been asserted that the United Kingdom was the leader of the Industrial Age, that the United States is the leader of the Information Age, and that China intends to be the leader in what some are calling the future Biotechnology Age. While this tidy way of describing technology development might be convenient, it fails to account for the global contributions in each of these different identified "ages." While the United Kingdom could be arguably identified as the birthplace of the Industrial Age, US industrial prowess contributed to growing our economy into the world's largest, with 25 percent of the world's wealth and only 4 percent of the global population. Such a formulation also indicates that the ages are distinct and can be accurately defined. A more useful concept might be to see these three ages as overlapping and converging to provide capabilities in each where the outcomes are far greater than the sum of the parts. One need only think of what lies at the convergence of 3-D printing (industrial), the Internet of Things (informational), and genetic synthesis (biological) to understand the potential.

Identifying the single authority or source for all technology development might be a laudable goal, but as our lexicon discussion should have reinforced, such a simplification is neither possible nor useful for understanding technology. One should expect that the global nature of technology development, multinational companies, and global societal trends will only continue to proliferate technologies throughout the world. We should also expect greater and more rapid advances in the individual technologies and combinations of technologies that will be created.

Examining funding trends can be useful in understanding the evolution of technology development. However, given our working definitions of science and technology; research and development; and innovation and transformation, it should now be obvious that discrete boundaries between these activities do not clearly exist, that a mix of formal and not-so-formal technology development activities are occurring, and that it is not entirely possible to delineate or fully capture all these efforts.

Spending on science is not normally reported and is certainly not accurately tracked globally. For example, inevitably important low-level scientific research and developmental activities being undertaken in small labs or emerging businesses would be overlooked. If one were to consider the data from the early 1970s, they would not include the funding associated with the early efforts of the Apple Corporation being developed by Steve Jobs and Steve Wozniak in a garage in Los Altos, California. On the other side of the ledger, technology development would likely not capture all the late-stage development that occurs in bringing a technology to market. Still, the data captured from statistical reporting sources provide an interesting point of departure for thinking about who is engaged in R&D, how much is being spent on these efforts, and toward what fields the R&D efforts are directed.

Unfortunately, the tracking and accounting systems prevent completely understanding who is doing technology development, what types of activities they are doing, what the technology priorities are, and what likely future trends will be. Those involved in government-sponsored programs might provide better definition and understanding of the activities being undertaken, as Congress appropriates resources based on categories such as operations and maintenance, R&D, and procurement; however, even with these delineations, the expenditures would likely be gross approximations, as many of these programs or activities likely get resources from multiple sources (the "pots of money" mentioned earlier). For industry, the data would likely be less rigorous, as each company's accounting systems and expenditures are unique. For large integrators that focus on government support, some standard definitions are likely in use. However, the S&T, R&D, and I&T being done in medium and small businesses and even in do-it-yourself industries is unknowable.

Despite these limitations, some estimates are available for individual components of S&T, R&D, and I&T. The National Science Board's *Science and Engineering Indicators 2018* report provides statistical data on R&D being conducted across the globe and notes several interesting trends, depicted in figure 2.1.[18]

To reinforce the point about the difficulty of capturing all the data, consider that the report relies on six major surveys of organizations that fund the "bulk" of all US R&D.[19] Some businesses or small, privately funded entrepreneurs may not want to release trade secrets or do not have a direct way for their R&D efforts to be captured, which contributes to significant underreporting as well.

Since 1953, when the United States began to track R&D spending, several important trends have emerged. First, approximately 54 percent of R&D annual spending was initially from government, while business accounted for 42 percent, and "others" (which included nonfederal government, higher education, and other nonprofit organizations) accounted for approximately 4 percent.

Figure 2.1. Research and development:
US trends and international comparisons[20]

- Worldwide R&D performance totaled an estimated $1.918 trillion in 2015, up from $1.415 trillion in 2010 and $722 billion in 2000. Fifteen countries or economies performed $20 billion or more of R&D in 2015, accounting for 85 percent of the global total. The top rankings at present continue to be dominated by the United States and China.
- R&D performed in the United States totaled $495.1 billion (current dollars) in 2015 and an estimated $475.4 billion in 2014. These numbers compare to US R&D totals of $433.6 billion in 2012 and $454.0 billion in 2013. In 2008, just ahead of the onset of the main economic effects of the national and international financial crises and the Great Recession, US R&D totaled $404.8 billion.
- The business sector continues to account for most of US R&D performance and funding.
- Most of US basic research is conducted at higher education institutions and is funded by the federal government. However, the largest share of US total R&D is experimental development, which is mainly performed by the business sector. The business sector also performs the majority of applied research. Although the absolute dollar values and actual shares have changed over time, these broad trends have remained mostly consistent for several decades.
- The business sector remains by far the largest performer in the US R&D system. R&D is performed across a wide range of manufacturing and nonmanufacturing sectors. R&D intensity is concentrated, however, in a few industries: chemicals manufacturing (particularly the pharmaceuticals industry); computer and electronic products manufacturing; transportation equipment manufacturing (particularly the automobile and aerospace industries); information (particularly the software publishing industry); and professional, scientific, and technical services (particularly the computer systems design and scientific R&D services industries).

By 1980, the federal government and business spending on R&D stood at approximately 47 percent, with approximately 6 percent in the "other" category. By 2015, the government had dropped to approximately 25 percent, and industry stood at approximately 68 percent, while the "other" category accounted for slightly less than 8 percent. In other words, there was an important shift in R&D expenditures during this sixty-year period.

The distribution of the R&D spending in terms of basic and applied research and experimental development also highlights relatively stable trends. In the United States in 1970, basic research accounted for almost 14 percent, applied research for approximately 22 percent, and experimental development for slightly over 64 percent. In 2015, spending in these categories was approximately 17 percent, almost 20 percent, and just over 63 percent, respectively, indicating the historical spending in these three categories has remained relatively consistent over the course of this time horizon.

Separating Department of Defense (DoD) from non-DoD R&D shows interesting trends as well. The DoD breakdown is basic research at 3 percent, applied research at 7 percent, advanced technology development at 8 percent, and major systems development at 81 percent. For the non-DoD component of R&D funding, basic research is at 44 percent, applied research at 41 percent, and development at 15 percent. The disparity is largely due to the major acquisition programs within the DoD that require advanced technology development and major systems development as part of the federal acquisition program. This is not to say that other departments and agencies do not also engage in these activities, but they spend far less on them.[21]

Another important trend since 1953 has been the growing importance of academic institutions, federally funded research and development centers (FFRDCs), and other nonprofits. In 2015, they accounted for 13.1 percent, 3.8 percent, and 4 percent, respectively, or almost 21 percent of all US R&D funding. Businesses accounted for 71.9 percent, and federal intramural—a term to describe internally conducted R&D—accounted for 7.2 percent. In fact, almost 63 percent of all basic research is done by academia (49.1 percent) and nonprofits (12.6 percent). The totals for academia and nonprofits go down considerably for applied research (24.3 percent) and experimental design (3.0 percent). These totals indicate the critical importance of academia and nonprofits in early research efforts.

For 2016, federal spending on R&D was $142.6 billion. The DoD accounted for nearly 50 percent of the spending at $69.1 billion. The other departments and agencies breakdown was:

Health and Human Services	$32.0 billion
Energy	$13.3 billion
National Aeronautics and Space Administration	$12.3 billion
National Science Foundation	$6.2 billion
Agriculture	$2.5 billion
Commerce	$1.9 billion
Transportation	$1.1 billion
Homeland Security	$886 million
Interior	$850 million
Veterans Affairs	$673 million
Environmental Protection Agency	$513 million

Organizations that received federal R&D funding of less than $500 million per year for 2016 were not listed.

For fiscal year 2017, the federal budget for R&D was projected to show an even greater disparity between defense and nondefense R&D. Defense R&D was projected to be $80.8 billion—a growth of $2.9 billion or 3.7 percent, while the nondefense R&D was reduced by $1 billion to $69.3 billion, a 1.5 percent reduction.[22]

Global diffusion of technology can be seen clearly in the statistics as well. While US R&D spending continues growth in real terms, other nations and regions have begun investing at rates comparable to or greater than US investment totals. In 2015, the world spent $1.9 trillion on R&D. North America accounted for $535 billion or 27.9

percent, Europe spent $415 billion or 21.6 percent of global R&D spending, and East and Southeast Asia as a region spent the most on global R&D at $721 billion or 37.6 percent of total.

In terms of total spending on R&D, the United States continues to lead. In this category, the top eight countries were:

United States	$496.6 billion
China	$408.8 billion
Japan	$170.0 billion
Germany	$114.8 billion
South Korea	$74.1 billion
France	$60.8 billion
India	$50.3 billion
United Kingdom	$46.3 billion

However, in terms of percent of GDP spent on R&D, the United States is no longer the world leader, spending only 2.7 percent of GDP. Other nations that spend more in this category include:

Israel	4.3 percent
South Korea	4.2 percent
Switzerland	3.4 percent
Japan	3.3 percent
Sweden	3.3 percent
Austria	3.1 percent
Taiwan	3.1 percent
Denmark	3.0 percent
Germany	2.9 percent
Finland	2.9 percent

By comparison, China and Russia only spend 2.1 percent and 1.1 percent of their GDP on R&D, respectively.

The overall conclusion from these trends is that as a percentage of global R&D, the US share is declining, indicating that technology is being democratized. While we still lead in terms of overall R&D spending, the rest of the world is catching up. The disproportionate amount of DoD R&D spending could have important impacts in non-DoD R&D as well.

TOWARD PURPOSEFUL TECHNOLOGY DEVELOPMENT

When I discussed the idea for this book with a colleague, he described a cutting-edge technology that had broad applications for identity management, screening and vetting, selection of training methods, interrogations, and candidate selection for certain professions (in this case military pilots). To make matters more interesting, the technology relied on competencies across a wide variety of technology areas, including optics, neurology, learning and development, biometrics, and big data analysis, to name a few.

The theory behind the technology was to determine the relationship between external stimuli and physiological reactions. Observing and recording reactions to certain stimuli was done with the goal of correlating stimuli and reactions to develop a means of predicting outcomes. By using a camera to look back through the subject's pupil to scan the eye muscles, the technology could record the muscle contractions, and the responses could be "read" to accurately predict certain behaviors, such as truthfulness versus deceit; the types of training that would likely be the most beneficial; or responses to external stimuli.

In describing this technology, my friend enthusiastically conveyed that the use case currently being contemplated was perfecting individual training methods for US Air Force pilots. It turns out that interest in this use case occurred largely by happenstance, as the developer of the technology was sitting by a USAF officer responsible for the training of new pilots at a conference. During a casual conversation on the edges of the conference, the tech developer told the officer about his concept, which aligned with the operational need to be able to select and train pilots.

While in this case, the tech developer and the operator have connected, a broader question arises of whether such randomness in marrying up the technological capabilities with operational problems is adequate for developing these connections. Asked another way, is a random process the most efficient way to ensure that the scarce resources associated with S&T, R&D, and I&T are most effectively employed? Or should (or can) a more deliberate process be developed to identify opportunities for convergence of technologies and eventual transition to use? More on this issue will be discussed in subsequent chapters, as it is central to our understanding of technological development.

CONCLUSIONS

In this chapter, we have identified the inherent difficulties with defining and deconflicting the various terms in the technology lexicon. In doing so, we have highlighted comparisons and contrasts between science and technology; research and development; and innovation and transformation.

Practitioners in technology fields of study will find that the terms discussed in this chapter are often used interchangeably. It is a shorthand that many have come to accept and understand. However, for the purist, this likely causes a great deal of discomfort. Scientists, technologists, and engineers in pursuit of knowledge are likely keenly aware of the tension between science and discovery to gain a fundamental understanding of natural phenomena and the use of scientific principles and knowledge for practical purposes.

In the next chapter, we will examine the foundations of technology development. We will also discuss the tension between near-term capability development and the longer-term, higher-risk search for fundamental science.

CHAPTER 3

FOUNDATIONS OF TECHNOLOGY DEVELOPMENT

This chapter examines the foundations of technology development. The starting point for the discussion will be definitions from the previous chapter. As the title of the book—*The Story of Technology*—indicates, our primary emphasis will be on technology development, while bringing in the related terms—science, research, development, innovation, and transformation—as appropriate to explain the evolution of technology and considerations for thinking about its path into the future.

This is not to minimize the importance of any of the other terms. Rather, we will examine those subsets of activities that directly pertain to technology development. The science from S&T, the research from R&D, and those elements of innovation and transformation directly related to technology will be considered as we go forward in our discussion. Given this overlap, we will undoubtedly, on occasion, delve back into each of the terms so as to facilitate discussion of our primary topic, technology development. And for our purposes, the focus will be on the purposeful development of capabilities, which throughout the chapter will be addressed as technology development.

As we consider technology development, and remembering that technology is the "application of capabilities for practical purposes," ask yourself: how does the technologist determine the practical applications for which capabilities are required?

After examining the structure of technologies, considering the properties that define technology development, and considering the rapidly increasing pace of technology development, we will arrive at an overarching theme: convergence.

TECHNOLOGY DEVELOPMENT AND THE S-CURVE

Since the late 1970s, the technology development S-curve has served two purposes. It provides a useful depiction of technology maturity and its rate of diffusion over time (see fig. 3.1), and it accomplishes this by providing a logical way to describe technology advancement by relating performance and effort required over time as a technology matures.[1]

The "S" sits within a graph in which the y axis measures performance and the x axis measures effort (or resources) required. The theory behind the S-curve is that it takes a great deal of early effort for small increases in performance. During the initial stages of technology development, progress would likely be measured in very small, incremental understandings or improvements. It might also be measured in a multitude of failures that serve as lessons that can eventually be applied to mastering the development of the technology. Over time and with continued effort (and assuming no one is attempting to violate fundamental laws of physics), the technology will mature and reach an inflection point at which the rate of change in performance begins to accelerate. During this exponential change, several occurrences will likely be observed.

First, less effort will be required to gain increases in performance. This does not imply that now is the time to stop providing resources to the technology development. Rather, it signifies that achieving increases in performance will be less costly in this phase of technology development. In other words, fewer resources will be required per unit of increase in performance. This is a good thing, as it means the technology is beginning to mature.

Second, progress along the S-curve will likely continue unless impeded in some way. Examples could include cases in which the technology is not feasible, other technologies may be required to support the development efforts, adequate resources are not available, or perhaps there is no demand for the technology. In cases where technology development is not progressing, perhaps additional research or scientific discovery will be needed. This is what is implied when we talk about taking a technology back to earlier stages of research. Alternatively, the development of the technology could depend on other component technologies that need to be either matured or integrated to support further development of the primary technology under consideration. It is also possible that there is little or no demand for the technology—that is, users are not looking to have a technology for the practical application envisioned.

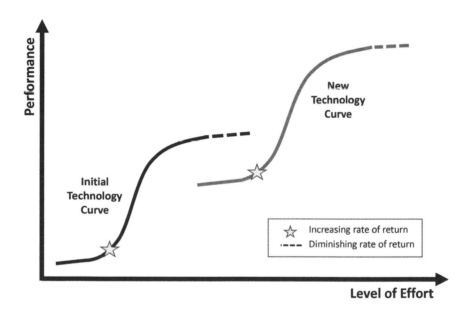

Figure 3.1. The Technology S-Curve

Third, as the technology matures, it will likely contribute to the understanding of other possible use cases. Initially, the technology may itself be developed as the primary application; however, secondary uses will inevitably be discovered. Technologies become building blocks for other technologies, with wide adoption expected for successful technologies. Consider how the basic technology of radio frequency and wave propagation, although many generations removed from its ancestors, has become an essential component in all wireless communications in use today.

Fourth, should specifics of the technology development become known or other similar technology development efforts occur in parallel, it is possible that a technology "pull" will take place. Such a demand signal could propel the development of a related or building-block technology. For example, the continuous product improvement that personal cellular telephones are undergoing demonstrates the symbiotic relationship between maturing technologies and increasing user demand. In the case of the cell phone, companies are adding capabilities to attract customers, and in turn, these new capabilities result in a new demand signal placed on the industry by consumers.

Another example of technology continuing to evolve in response to a practical need is gene editing. While gene editing has been around since the 1970s, using techniques such as zinc finger nucleases and transcription activator-like effector nucleases, these methods required a great deal of skill and knowledge and were not particularly rapid or accurate. The development of CRISPR-Cas9 essentially changed the world of gene editing by developing commercially available kits that provide speed, precision, and accuracy to manipulate a genomic sequence. The success of CRISPR is leading to the development of more gene editing tools with even greater capabilities.[2]

Fifth, a technology will reach the point of diminishing returns. Over time, when a technology is reaching full maturity, the rate of increase in performance will begin to slow. In our S-curve model, the cost for incremental improvements in performance becomes prohibitive. This would likely be the point at which the prudent technologist and his investor decide to withhold additional resources for further technology development.

It is also at this point in the technology life cycle that several valuations of the technology are possible. The technology could be determined to be mature and not likely to benefit from further development, meaning it will not be updated for use in newer models or versions. It could become widely adopted for use as a subcomponent in other primary technologies. Or it could be determined to be obsolete and allowed to reach the end of its useful life cycle.

Sixth, in the technology progression, one might ask whether the original S-curve has reached a terminal point and a new S-curve should be started. For example, over the last forty years, the state of the art in computing has seen a steady progression of platforms and capability: mainframes, minicomputers, the introduction of personal computers, mobile computer processing, and now cloud computing as part of the mobile web and the Internet of Things (IoT). The transition from the fixed computing of the mainframes, minicomputers, and personal computers to mobile computing, the cloud, and the IoT could serve as such a delineation.

Continuing with this example, one can also think of the nature of cellular telephone communications. In the early 1980s, the early cellular systems and their embedded technology went through a series of developmental stages. The first S-curve can be thought of as 1G capability, which started at about 9.6 kilobytes per second. In the early 1990s, technology development had led to the establishment of a 2G CMDA network with speeds starting at approximately 23.8 kilobytes per second. In the early 2000s, another generation of mobile communications began to be fielded. The 3G network had speeds starting at approximately 64 kilobytes per second. Finally, around 2010, 4G networks were being fielded that had speeds over 384 kilobytes per second. Each of these nested S-curves represent a new generation of cellular technology. The fielding of 5G will usher in the next generation of mobile devices, bringing high data rates, reduced latency, and massive data connectivity that will support the demands of the IoT.

S-curves can also be very useful for understanding the rate of technology adoption. The categories describing this rate are innovators, early adopters, early majority, late majority, and laggards. Obviously, if one is interested in technology from the standpoint of investing, being an innovator or early adopter is likely to have the most benefit. On the other hand, if one is thinking about this as a consumer, innovators and early adopters are likely to get products with less maturity, fewer capabilities, and perhaps some flaws for the price of early entry into the market. At the other end of the spectrum, late-majority consumers and laggards are likely to get more mature technology but may be entering the market when newer, more advanced products are already becoming available.

Many factors are likely to affect one's preference for when to enter the market. If one is very risk tolerant, early adoption could make sense. However, the risk averse would likely be far less comfortable with a technology that has not reached maturity, likely has undiscovered flaws, and will occasionally not operate as intended (i.e., fail).

Previously, we discussed technology development as not being a linear process. Certainly, the very shape of the S-curve implies non-linearity, with performance increasing slowly initially, followed by an inflection point and an exponential rise in performance before tailing off once again as the technology reaches maturity. But there is another way in which the technology development is nonlinear. In some instances, it will be necessary to return to earlier stages—or earlier technology readiness levels (TRLs)—should a technology or any of its component technologies not be ready. It is entirely possible to believe the R&D has been completed to allow for moving the technology to a higher TRL, only to discover that additional R&D will be required. Perhaps in progressing in development, it becomes obvious that certain fundamental understandings are lacking and therefore additional scientific discovery or research must be conducted.

Intended or not, the technology readiness levels and S-curve concepts have become intertwined. One can easily envision that TRLs 1–3, representing basic and applied research, can be on the lower part of the S-curve before the first inflection point; that technology development represented by TRLs 4–7 coincides with the center part of the S-curve; and finally, that TRLs 8–9 occur after the second inflection point. While the concepts were designed for different purposes—in the case of the S-curve, to describe how technology matures, and in the case of TRLs, as part of a research, development, and acquisition life cycle—the overlaps are clear.

HOW TECHNOLOGIES ARE STRUCTURED

In his book *The Nature of Technology*, Brian Arthur does a great service to the field of evolutionary technology, providing an important perspective regarding principles of technology development. It is through these understandings of history and the evolutionary principles that one gains not only an appreciation for both technology development in the past and the trajectory of technology today, but also a view—perhaps only a peek—into the future of technology.

Arthur's view of technology rests on three core principles:

1. All technologies are combinations of elements.
2. These elements themselves are technologies.
3. All technologies use phenomena to some purpose.

The field of aviation provides an interesting example of Arthur's principles applied to a technology. Humankind's interest in flight had been well documented over the past 2,000 years. Humans' attempts to emulate birds in flight by jumping off towers covered in bird feathers met with a predictable fate. The Chinese developed an understanding that hot air rises and therefore could be used to provide lift assuming the proper materials could be found.

In the 1500s, Leonardo da Vinci wrote the *Codice sul volo degli uccelli* (*Codex on the Flight of Birds*), which detailed in thirty-five thousand words and five hundred sketches the nature of air, bird flight, and concepts for developing flying machines.[3] Da Vinci's work often related to solving military problems, from weapons design to aerial reconnaissance. The latter reflected his interest in the nature of flight.

In the 1780s, hot-air balloons were developed and demonstrated sustained flight. Interest in aviation grew in the early 1800s as aviation pioneers, now armed with greater technical training, began to incorporate these ideas toward human flight possibilities. Concepts for heavier-than-air flight began to be understood, relying on theories and observations of birds in flight.

Here in the United States in the late 1890s, Orville and Wilbur Wright began developing concepts for manned, powered flight. Their work was based on three foundations: their knowledge obtained as mechanical engineers and inventors, the work of other aviation pioneers, and observation of natural phenomena. Their first aircraft was a small biplane kite structure with two wings and measuring only five feet across. It incorporated basic aerodynamical principles, using the wings to generate lift that would provide the force to allow the structure to remain airborne. They developed a homemade wind tunnel, conducted glider tests, and analyzed the data to improve their understanding of aerodynamics. These tests also enabled them to develop the engines that would provide the necessary speed to generate the lift required for powered human flight.

Their independent observations of birds in flight provided the basis for solving one of the difficult problems that previous aviation pioneers had encountered, the control of the aircraft. They invented the concept of three-axis control to steer the aircraft effectively and to maintain its equilibrium during flight. In fact, it is interesting that their first patent was not for the invention of powered human flight but rather for technology associated with aerodynamic control.[4]

The very same technologies and natural phenomena used by the Wright brothers to overcome the challenges of early powered human flight are now employed in all modern aircraft. Obviously, aviation technology has matured, and the number of subsystems and components in modern aircraft has increased greatly from those early days of flight. Still, the parallels between early aviation pioneers and today's aviation systems remain.

The Wright brothers were concerned with the integration of the wings that provided the lift, the engine that provided the propulsion, and the control surfaces that allowed for steering the aircraft. The wing designs of today are direct descendants from those early days. The small propeller-driven engine has been replaced by large jet engines used individually or in combination to provide thrust for supersonic aircraft flight. The control systems have been greatly matured and augmented with features that provide stability for aircraft that are inherently unstable and are therefore flown by computers rather than humans.

Modern aircraft include numerous other technologies that have been combined, integrated, and synchronized into the aviation systems of today. As Brian Arthur reminds us, each technology is the combination of other technologies. Human flight today, regardless of the intended purpose, includes wings, engines, and control systems, all wrapped around a climate-controlled fuselage that enables us to survive at high altitudes where oxygen is virtually nonexistent while traveling at extraordinary speeds that allow us to cover great distances in a short period of time. However, this does not tell the entire story. For example, the materials that make up the wings and fuselage have been developed based on technologies for metallurgy, materials science, and lightweight aircraft structures. Advances in materials science have led to the use of composites in place of metals, making structures lighter and increasing the efficiency of the aircraft. This substitution can be seen in Boeing's 787 aircraft, which takes great advantage of these new materials.

The control systems used in modern air flight include electronic sensors that provide information to the pilots and flight computers. Autopilot controls enable aircraft to essentially fly themselves based on preprogrammed flight plans.

So far, we've been focusing on the aircraft as the major technology and using the concept of the wings, engines, control systems, and structures as the subsystems. It is also possible to look more broadly at aviation and see that the field has been greatly aided by complementary increases in other systems and technologies. Advances in aviation have spawned the development of commercial airline travel for humans and cargo. The incorporation of new technologies has made airline travel more effective, efficient, and safe. Airports have been built to support airline travel and a global aviation industry. Runways provide safe landing strips for aircraft. Air traffic controllers direct the aircraft pilots on final approach to the airfield. A larger system of air traffic controllers essentially develops highways in the sky that deconflict the paths of aircraft to minimize the chance of midair collision.

As airline travel has become routine and efficient, industries have turned to aircraft for delivering commodities that must be rapidly brought to market. Flowers from Colombia in Latin America are shipped routinely to destinations in the United States. Perishable or time-sensitive goods can be rapidly shipped to destinations where consumers eagerly await their arrival. Global supply chains, which include aircraft used to ship goods around the world, are fueled by powerful technologies—methods and algorithms—that calculate the most efficient ways to transport time-sensitive products.

Security and defense have also benefited from widespread incorporation of aviation applications. The early stages of Operation Desert Storm, from January 17 to February 23, 1991, were largely a coordinated air campaign that incorporated manned and unmanned aircraft and missiles employing precision-guided munitions with great effect and destroying key facilities and troop formations. Patrolling and observation along the southwest border of the United States have been augmented with unmanned aerial systems and aerostats to stem the flow of illegal immigration and illicit narcotics. Aviation platforms are also regularly employed for missions ranging from surveillance and reconnaissance to delivery of precision munitions. Law enforcement, first responders, and the US Coast Guard routinely employ aviation for medical evacuation and search and rescue.

Advances in aviation have also led to complementary advances in other areas. Wind tunnel science has been developed to improve aerodynamics for aircraft. However, it also has a secondary benefit for use in improving the aerodynamics of automobiles, racing bicycles, and any other conveyance that places a premium on decreasing aerodynamic drag. In the future, Amazon is planning to use small unmanned aerial systems or drones for the delivery of merchandise and groceries.

In this way, we come to see that it is not the individual technology alone that matters but the way technologies are combined into systems that are part of other systems designed to achieve practical purposes. This hierarchy from concepts to individual technologies to technology systems is not the exception but rather the rule. It can be observed across any and all fields of technology.

The field of computing provides another illustration. Early numbers theory directly contributed to the field of computers, from the abacus invented in Babylon in 2700–2300 BCE to the latest high-performance computing systems, which rely on massively parallel processing systems measured in teraflops (or 10^{12} floating point operations) per second. This lineage includes early mechanical devices for calculating astronomical positions and cryptography; the work of Alan Turing in developing theoretical computer science and the Turing machine; the 1946 Electronic Numerical Integrator and Computer (ENIAC), which was the first publicly available general-purpose electronic computer; the early internet and packet switching technology; and today's wide variety of mobile digital computers, such as smartphones and the embedded computing technologies that are building blocks of the IoT. As we look to the future, we can clearly see that the building blocks of AI can also be observed in these early advances in the field of computers. This lineage has also provided foundations for other, perhaps seemingly less related fields, such as modern finance and banking, transportation and navigation, space exploration, and weather prediction, to name a few.

It is through the science and technology, research and development, and innovation and transformation, which we have been loosely referring to as technology development, that the integration, coordination, and synchronization of the systems of systems actually occurs. In fact, our definition of technology as "the application of capabilities for practical purposes, which can include inputs from scientific knowledge or the collection of techniques, methods, or processes applied in a purposeful way" nicely describes these systems of systems we encounter in our daily lives.

CONNECTEDNESS VERSUS DISCONNECTEDNESS

A classic dilemma in technology development is the question of connectedness versus disconnectedness. Something is said to be *connected* if the technology development directly relates to an operational need and *disconnected* if there is either no relationship or a loose relationship between the technology and the operational need.

Connectedness occurs when technology developers are working closely with operators to identify real-world problems and look for solutions. Disconnectedness, on the other hand, occurs when scientists, researchers, and developers are largely working independently without deep understanding of the operational problem that is to be solved. One can easily see how either of the two extremes—either absolute connectedness or complete disconnectedness—would not be beneficial.

In the connected world of technology development, longer-term opportunities undoubtedly would be sacrificed for the benefit of near-term solutions that could be helpful immediately. When Henry Ford was asked about innovation and responding to customer requirements, he said, "If I had asked people what they wanted, they would have said faster horses."[5] They certainly would not have been able to describe the car, which has become ubiquitous as a mode of travel today. This represents the classic problem for technology development.

The customer understands what would make her life easier and therefore, when asked, describes satisfying her near-term capability requirement. Perhaps the issue is connecting a computer to a printer, and the request is to develop twenty-foot-long cords that would allow for connecting computers and printers across a room. However, what if the solution lies not in a longer cord but in the use of smart hubs placed every five feet within the distance of a computer that would establish a network for connecting the computers? Or, thinking further into the future, perhaps the solution lies in developing a wireless network that allows for establishing a local area network. The customer only knows that the computer and printer need to be connected but is perhaps unaware of what technology solutions might be available or could be developed.

If the developer is too close to the problem, only near-term solutions or incremental improvements are likely. Thus, with the rate of change, a technology or "solution" might well be delivered just as it is becoming obsolete. In other words, something in the environment might have occurred to make that technology less useful. Some might even say the newly delivered technology is solving yesterday's problems.

When we examine disconnected technology development, an equally disconcerting problem rises. Technology developers understand what the future possibilities hold, and perhaps they develop capabilities and systems, thinking these might be beneficial based on their technologically informed perspective. However, this perspective likely does not include an operational viewpoint. If the technology developer does not understand the operational problems or only has a vague conception of them, the developer is beginning by violating the spirit of technology development.

Potentially, the result could be a very interesting gadget with no practical application. This would amount to a waste of resources if the technology has no practical purpose. On occasion, perhaps the developer will be lucky and a practical application will be identified. However, this seems to be a risky proposition, given that the funding for research and development is scarce. As a result, if one wants to get the most out of each R&D dollar, alignment with an operational outcome and the operator becomes a necessity.

So what is the answer to this vexing problem?

The roots of this connectedness-disconnectedness question go back to the post–World War II period. Warfare has always brought out the technology and innovation in humankind. Early warfare saw the development of spears and swords. World War I saw the introduction of tanks, flamethrowers, poison gas, air traffic control, and aircraft carriers.[6]

World War II was no different. Advances included the development and introduction of the first computer, jet engines, V-2 rockets, radar, penicillin, radio navigation, and, of course, nuclear weapons.[7] It is not an overstatement to say that the rate of change of technological development was heavily influenced by the global and national emergency of World War II. For example, the British were already developing radar, but their sharing of that technology with the United States, and its rapid inclusion in weapon systems, would not have happened in the same time frame had it not been for the necessity of integrating it into Allied weapons.

No World War II technological development process was more catalyzing than the creation of nuclear weapons. World War II created an
urgency for developing atomic weapons that would not have existed in
the absence of the war. Understanding the natural phenomena related
to a wide variety of technical specialties, including preparation and
handling of atomic material, sustaining a nuclear chain reaction, and
developing specialized triggers for detonation, was essential for the
Manhattan Project. Each required some amount of basic and applied
research and was essential for developing a nuclear weapons capability.

In terms of the expansiveness of this development effort and its
overwhelming battlefield effect, nuclear weapons and their development represented an inflection point in the US role in global affairs,
our national security, and the concepts that governed the development
of technology. Many of the lessons learned from World War II and its
immediate aftermath can be directly traced to our modern national
security and R&D processes. And one of the most immediate concerns
was the connectedness-disconnectedness question surrounding technology development.

For obvious reasons, during World War II research and development was clearly connected to operational requirements. Operational
imperatives drove operational requirements that drove focused R&D.
There was little time for basic science and gaining insights into natural
phenomena other than to satisfy operational concerns.

It was also during the war years that the age of modern R&D
thinking began to emerge. One of the people who was instrumental in
technology development during the war effort was Vannevar Bush, who
headed the US Office of Scientific Research and Development, which
oversaw almost all wartime military R&D including the initiation of the
Manhattan Project.[8] Through his relationship with President Franklin
Roosevelt, Bush had a disproportionately strong voice toward the end
of the war in shaping the postwar science and technology enterprise.

In response to four questions from President Roosevelt, Bush prepared what became an enormously influential report, *Science, the Endless
Frontier*. In it, he argued strongly for the importance of maintaining

independent science research, divorced from the bindings of operational requirements to allow for continued discovery.

His wartime experiences had convinced Bush that operational necessity would always detract from longer-term basic and applied research. In short, he believed that the tyranny of the here and now would overtake the quest for longer-term scientific discovery. This argument began to frame the discussion between connectedness and disconnectedness that in many respects continues today. More on Vannevar Bush and the Manhattan Project will be discussed later, as both are critically important to the evolution of modern R&D in the United States.

In a book titled *Pasteur's Quadrant: Basic Science and Technological Innovation*, author Donald Stokes provides a framework for illustrating this connectedness-disconnectedness problem. The impetus for this book derived from concerns he had about Vannevar Bush's expressed desire to preserve resources for pure science. *Pasteur's Quadrant* takes a more nuanced view regarding the need to maintain balance between connectedness and disconnectedness in R&D.

Stokes develops a matrix consisting of four quadrants that are formed by considering whether the technology development—that is, the basic science and technological innovation—is being conducted as a high or low quest for fundamental understanding and whether there is a high or low consideration for use. The lower-left quadrant has both a low quest for fundamental understanding and a low consideration of use, meaning that there is no reason for scientific research to be conducted for these types of questions.

The top-left quadrant comes at the intersection of a high quest for fundamental understanding and a low consideration of use. Stokes refers to this part of the matrix as the "Bohr quadrant," named after Niels Bohr, a physicist who received the Nobel Prize for his work on the structure of the atom. Bohr's studies tended toward being more theoretical in nature and, in addition to his work on atomic structure, included such subjects as surface tension caused by oscillating jets, fundamental inquiries into radioactive phenomena, absorption of alpha rays, and quantum theory. While his work was considered highly theoretical in its day, it formed the intellectual underpinnings and foundations for a wide range of scientific principles, technologies, and inventions that have become integral to humankind, such as the structure of the atom, quantum physics, and quantum computing. Stokes identifies this quadrant as the realm of pure basic research.[9]

The lower-left quadrant comes at the intersection of a low quest for fundamental understanding and a high consideration of use. Stokes labels this part of the matrix the "Edison quadrant," named after Thomas Edison, who was largely recognized as an American inventor. Edison is credited with holding 1093 patents; establishing the first industrial research laboratory; and creating a host of specific inventions and innovations related to the telegraph as well as early batteries, the telephone, the microphone, incandescent lamps, and a motion picture projector. In a very real sense, he was more of an inventor or perhaps an engineer than a scientist. He was constantly looking for the practical applications for technologies to solve real-world problems. And he was clearly not in the theoretical world of Niels Bohr. Stokes identified the activity in this part of the matrix as applied research.

The final quadrant in the top right was where there was both high quest for understanding and high consideration of use for the research. This quadrant, labeled the "Pasteur quadrant," is the realm of use-inspired basic research. Pasteur was both a scientist and inventor. He was a research chemist who worked in a laboratory to gain fundamental understanding and solve problems of the day. He helped develop theories of how disease was spread as well as developing processes to remove bacteria from milk using a technique that now bears his name, pasteurization. Pasteur followed the scientific method in gaining an understanding of rabies and development of a new vaccine to protect against the virus.[10]

The question of connectedness versus disconnectedness in R&D continues to this day. The statistics on R&D spending presented earlier indicate that in 2015, basic research accounted for almost 17 percent and applied research accounted for almost 20 percent of the total R&D spending. Assessing whether this reflects the proper spending on basic and applied research will not be directly addressed, but it is an interesting question nonetheless. The answer is both complex and variable in that it will vary between technology fields and over time.

However, the connectedness-disconnectedness issue remains an interesting question and a point of consideration for future technology spending. How much spending should be focused on near-term issues and technologies for which there is a use case, versus for basic science necessary for gaining fundamental knowledge and for which there is no immediate use case, will likely remain a matter of important debate. Returning briefly back to Stokes's *Pasteur's Quadrant*, consider that the highly theoretical work of Bohr has now been translated into practical applications. Could these have occurred had operational needs prevented the more theoretical questions from being addressed?

TRENDS IN TECHNOLOGY DEVELOPMENT: THE THREE D'S (DUAL USE, DISRUPTIVE TECHNOLOGY, AND DEMOCRATIZATION OF TECHNOLOGY)

Three historical trends have defined technological advancement: dual use, the disruptive nature of technology, and the democratization of technology. While one can find ample evidence of each throughout the ages, what is new is the speed, global reach, and wide variety of uses of today's cutting-edge technologies. Understanding these trends provides a foundation for considering the challenges and opportunities that will likely continue to grow, bringing greater and more powerful technologies for the betterment of society—as well as unintended consequences, powerful tools for malevolent actors, and perhaps even unsettling societal stresses.

Dual Use

The term *dual use* has come to have multiple meanings. In casual conversation, we say an item is dual-use technology if it can satisfy more than one purpose at any given time. A technology is also said to be dual use if it has applications for commercial or civil purposes but also the potential to be used for military purposes (as weapons) or malicious applications. It is this second connotation that will be a primary point of discussion for us.

Specific definitions of dual use are contained in US law and controlled by two departments within the government, the Departments of Commerce and State. These provisions will be in discussed in greater detail in the chapter on managing technology, as they form the basis for US export control efforts.

Our goal for introducing the concept of dual use at this point is to provide a foundation for understanding how technology can be employed for competing purposes, as both a benefit and danger to humankind. An example assists in explaining this dual-use conundrum.

In testimony to Congress, a former commander of the US Army Medical Research Institute of Infectious Diseases used the development of a vaccine versus a biological weapon as an example of a dual-use problem. He illustrated the issue by highlighting that the first four steps in both processes are identical: isolation of the pathogen, engineering the biochemical properties, growing the biological material, and using an animal model to test the pathogen. The next five steps vary depending on whether one is developing a vaccine or a weapon.

For a vaccine, the pathogen would be attenuated, small quantities would be grown in the lab, preclinical trials would be conducted, and clinical trials would follow. Once the safety and efficacy of the vaccine were established, it could be certified for use in humans. If one were developing a biological weapon, the pathogen would be made more virulent and stabilized for deployment in the environment, dissemination methods would be developed, large quantities would be produced, and a weapons delivery system would be developed.

In the first case, the development of a vaccine, the activities would be permitted by the Biological Weapons Convention, the treaty that serves as the international norm against use of biological weapons, while in the second case, the activity would clearly be a violation of the BWC. As this example illustrates, the question of whether some scientific or research application is being done for legitimate or malicious purposes can be challenging to ascertain.

Earlier, the assertion was made that all technology is dual use regardless of the original design or intended purpose, as any technology could have application either for legitimate peaceful purposes or for dangerous illicit activity. Can a technology or, in this case, an experiment have utility for legitimate peaceful purposes and pose a dangerous illicit threat? This was essentially the question that confronted the US government in 2012.

Unbeknownst to many in the US government, two studies had been sponsored by the National Institutes of Health to examine whether the H5N1 highly pathogenic avian influenza (HPAI) could be made more transmissible while retaining its virulence. HPAI has a mortality rate of approximately 60 percent yet is not highly adapted to the epithelial linings of the human lung. In fact, most infections to date have resulted from close contact with blood from an infected avian. As a result, it is not highly transmissible by the aerosol route, which is a very good thing for humankind. Imagine the danger of a disease with the transmissibility of seasonal influenza combined with extremely high virulence.

Two separate research teams—led by Ron Fouchier from the Erasmus Medical Center in the Netherlands and Yoshihiro Kawaoka from the University of Wisconsin—were sponsored by the US government to answer this question.[11] The teams introduced mutations in the virus and passed it successively through ferrets to determine if the virus would adapt to the aerosol route while retaining its virulence.[12] The teams were not successful in creating a super H5N1 strain that was both highly pathogenic and transmissible. Again, this is good news for humans.

Perhaps equally noteworthy was the government's response upon discovering that results of the two studies were going to be published. The government, as the sponsor for this research, stopped the publication of the articles until a thorough assessment could be made as to the potential harm of the release of the information. This kicked off a two-year process of discovery and analysis, during which the concept of dual-use research of concern, or DURC, became a hotly debated topic. The process provided interesting insights into the question of dual use.

During interagency deliberations, questions concerning the necessity of conducting potentially dangerous experiments were hotly debated. The issue centered on what became known as gain-of-function research—that is, increasing the functionality of certain attributes of a pathogen. Through this process, DURC would come to be defined as

> life sciences research that, based on current understanding, can be reasonably anticipated to provide knowledge, information, products, or technologies that could be directly misapplied to pose a significant threat with broad potential consequences to public health and safety, agricultural crops and other plants, animals, the environment, materiel, or national security. The United States Government's oversight of DURC is aimed at preserving the benefits of life sciences research while minimizing the risk of misuse of the knowledge, information, products, or technologies provided by such research.[13]

The DURC policy goes on to list seven experiments of concern and fifteen pathogens to be governed under it. The fifteen pathogens are largely those that have historically included bacteria, viruses, or derived biological toxins that have been thought to have the greatest potential to be used as biological weapons, as well as the H5N1 HPAI and the reconstructed 1918 influenza virus. The seven categories of experiments include any activity that

1. Enhances the harmful consequences of the agent or toxin
2. Disrupts immunity or the effectiveness of an immunization against the agent or toxin without clinical and/or agricultural justification
3. Confers to the agent or toxin resistance to clinically and/or agriculturally useful prophylactic or therapeutic interventions against that agent or toxin or facilitates their ability to evade detection methodologies
4. Increases the stability, transmissibility, or the ability to disseminate the agent or toxin
5. Alters the host range or tropism of the agent or toxin
6. Enhances the susceptibility of a host population to the agent or toxin

7. Generates or reconstitutes an eradicated or extinct agent or toxin listed.[14]

Several key thoughts are worthy of consideration in evaluating this DURC policy. The Fouchier and Kawaoka experiments were undertaken for the purpose of increasing the understanding of a natural phenomenon, in this case the spread of disease. They were conducted using proper containment facilities and employing the scientific method. These were not random experiments but, rather, employed well-considered analytical methods to determine the threat potential to humankind should the H5N1 virus mutate and become more adaptable to human epithelial lung tissue. In other words, they conformed to the principles embodied in the scientific method.

A couple of postscripts are in order for evaluating the DURC policy. As written, the policy pertains to a very small subset of science that would be sponsored by the US government, meaning that privately funded gain-of-function research could continue to be conducted. Second, several years later, Dr. Robin Robinson, the director of the Biomedical Advanced Research and Development Authority—the organization responsible for the development and purchase of vaccines, drugs, therapies, and diagnostic tools for public health emergencies[15]—offered that the H5N1 experiments conducted by the Fouchier and Kawaoka teams had provided invaluable information regarding the spread of influenza that would contribute to the improvement of the efficacy of the annual seasonal influenza vaccines.

Even prior to development of the DURC policy, the life sciences community had well-established codes of behavior governing R&D. These codes had been created due to the sensitive nature of work done in this field. Ultimately, R&D conducted in the life sciences is conducted for the purpose of using the information, knowledge, and treatments on living organisms. As a result, the requirements for safety and efficacy are quite high.

At the end of the day, mitigating dual-use concerns comes down to managing risk. As all technologies have the potential to be used for useful practical applications as well as for nefarious purposes, a framework and tools for managing these risks are necessary and are provided by laws, policies, regulations, and standards. It is also interesting to consider whether the seven experiments of concern could be adapted and might have applications in other fields beyond biology as limits in technologies such as artificial intelligence are considered. More on how technologies are "managed" will be discussed in a later chapter.

At times, dual use has been regarded from an unhealthy perspective. If one says something is dual use, the antenna goes up with immediate suspicion that a technology could be dangerous, perhaps even something to be avoided. But that is not always the case. The use of GPS technology provides such an example. GPS was originally designed for navigation of military assets and for controlling long-range weapon systems and other smart weapons. The same technology now forms the backbone of numerous civilian applications, such as navigation in all forms of transportation including aircraft, ships and automobiles. This simple example serves to illustrate that dual-use technology does not refer to something that is inherently bad or dangerous but, rather, that the use case—whether it is used for the betterment of society or for malicious purposes—should become the determinant for considering and enforcing dual-use concerns.

Historically, controlling the proliferation of technology, especially dual-use technology, has met with varying degrees of success. Suffice to say that national policies and laws in concert with coalitions of like-minded friends and allies have long sought to control the spread of dangerous technologies. More on this subject—including specific strategies—will be discussed in our chapter on managing technology,

Disruptive Technologies

Harvard Business School professor Clayton M. Christensen first used the term *disruptive technology* in his 1997 book *The Innovator's Dilemma*. Christensen separated emerging technologies into two categories: sustaining and disruptive. Sustaining technology relied on incremental improvements to an already-established technology, while a disruptive technology was one that displaced an established technology. A disruptive technology also had the potential to transform an industry or develop a novel product to create a completely new industry.[16]

In considering the question of disruptive technology, history provides numerous interesting examples. One such example is the printing press, which allowed humans to record and share information in an efficient and effective manner. Other disruptive technologies include the railroad and highway systems in the United States, which promoted commerce and travel across the nation. A more recent disruptive technology is represented by the recent IT revolution, which has resulted in development of a new domain called *cyberspace*. Each of these technologies has been disruptive to the status quo. And in each of these cases, the introduction of these technologies has resulted in an inflection point or even a discontinuity that has fundamentally changed the trajectory of society and altered the course of history.

Defining whether a technology is disruptive is not an exact science, but rather largely open to interpretation. In assessing whether a technology is considered disruptive, one study proposed four factors that could be used in making this determination: (1) the technology is rapidly advancing or experiencing breakthroughs, (2) the potential scope of the impact is broad, (3) significant economic value could be affected, and (4) economic impact is potentially broad.[17]

This same study went on to list twelve technologies that were considered disruptive. They included mobile internet, automation of knowledge work, Internet of Things, cloud computing, advanced robotics, autonomous vehicles, next-generation genomics, energy storage, 3-D printing, advanced materials, advanced oil and gas exploration, and renewable energy. It also highlighted several important technologies that were potentially disruptive that did not make the list, including fusion power, carbon sequestration, advanced water purification, and quantum computing.[18] Another such list of disruptive technologies retrospectively identified the personal computer, Windows operating systems, email, cell phones, mobile computing, smartphones, cloud computing, and social networking as disruptive technologies.[19]

As one might guess, others have developed different lists of potentially disruptive technologies that are on the horizon. It would likely come as no surprise that these multiple futurists' lists do not match one another, although many of the same technologies have been identified. This should not necessarily be troubling, as making such determinations relies on a mix of qualitative and quantitative judgments and is best understood looking through the lens of history rather than projecting into the future. In short, to truly determine whether a technology has been disruptive requires retrospectively examining its impact on humankind and society.

Making such predictions becomes even more challenging when attempting to do so looking at a single technology area. Rather in considering the degree to which a technology has been disruptive, one should examine not just the primary application but also the secondary adoptions of the technology that have occurred. For example, if one considered information technology in isolation, its impact on biotechnology and associated development of a new field called bioinformatics would likely not be evident. Nor would looking at information technologies in isolation have been useful for understanding how the combinations of these technologies are contributing to the growing advances in AI.

Likewise, it is not possible to look at biotechnology—say, gene editing—in isolation and project its trajectory or the extent to which it will become disruptive. Rather, one must look at related technologies or other enabling technologies that could be combined in ways that would be useful in developing new gene editing techniques. Based on changes being observed in gene editing technologies today, one can surmise that advances in information technology or bioinformatics, nanotechnology, advanced manufacturing, and neuropharmacology could play a significant role in plotting this trajectory. In fact, advances in gene editing could also be aided by technologies that do not currently exist.

Disruptive technologies can be very discomforting for individuals and society. The very definition of *disruptive* certainly implies a significant change that may place stresses and strains on the current human and technological systems that are in operation today. For this reason, it is important that the introduction of disruptive technologies at least consider and perhaps even seek to mitigate some of the deleterious effects.

Consider the societal implications and unease resulting from changes that are occurring at the intersection of technology and labor. A restructuring of the market space is likely to occur, resulting in the elimination of repetitive, low-skilled positions and placing a higher premium on positions relying on science, technology, engineering, and mathematics. This labor-market restructuring will occur based on advances in AI, autonomy, and robotics, aided by the synergistic effects of the IoT, 3-D printing, and big data science. The development of these technologies potentially brings the promise of higher productivity (and, with it, economic growth), increased efficiencies, safety, and convenience. However, this market restructuring also raises difficult questions about the broader impact of automation on jobs, skills, wages, and the nature of work itself.[20]

This is not to imply that technology development for disruptive technologies should not be undertaken, but rather that such technologies could have societal, organizational, psychological, moral, and ethical components that are far too important to not appropriately consider. One need only contemplate the uncertainty and concerns surrounding technology and the labor market to gain appreciation for how disruptive technologies can be to the fabric of society.

Democratization of Technology

The democratization of technology refers to the increasing accessibility of technologies across the world for individuals, industries, and nations. Through democratization of technology, more people have access to sophisticated capabilities, tools, and techniques that were once the exclusive purview of only a few. History provides limitless examples of the technology democratization that has been occurring, which is both accompanying and fueling the globalization that has been undergoing exponential growth over the past several millennia.

Today, the democratization of technology continues at an overwhelming pace. One need only consider the proliferation of cellular telephone capabilities to see an example of this democratization, where numerous generations of cellular technology have been developed within a twenty-five-year period. Looking beyond the dramatic increase in capabilities, where cellular technology has gone from voice-only platforms to mobile computers, one also sees the even more dramatic increases in adoption of the technology across the globe.

Early cellular telephones were large, clunky, and only accessible to a very few when they were introduced in the early 1990s. For communication in austere conditions, one was forced to rely on early mobile telephone technologies such as Inmarsat, which was initially used in a maritime environment for ships at sea.[21] Voice-only cellular phones began to see more widespread introduction in the late 1990s. While the history is not precise as to the exact point when smartphones were first developed, widespread use of mobile phones with voice, data, and video began to accelerate in the mid-2000s. With the introduction of the iPhone by Steve Jobs in 2007, it became clear that the smartphone generation had begun.[22] The proliferation of the smartphone continues to experience a rapid exponential climb. One report states that 70 percent of people in the world will be using smartphones and 90 percent of the world will be covered by mobile broadband networks by 2020. This means that more than 6.1 billion people will have access to smartphone technologies by 2020.[23]

Returning to Brian Arthur's concept of technologies being combinations of other technologies, we see that dual-use technology is no different. Therefore, as dual-use technologies mature and proliferate, they too will suggest other use cases. The GPS, developed for use in military applications, provides such an example. Today, it is ubiquitous in virtually the full range of military activities from maneuvering of forces to the gathering of intelligence. But it will also be available to the over 6.1 billion smartphone users in 2020.

The growing sharing of information will only serve to further promote the democratization of technology, making it more readily available for all. Consider that the lessons learned during combat operations in Iraq and Afghanistan regarding urgent medical care have been brought back into the emergency rooms of hospitals here in the United States. Through the knowledge and experiences gained on the battlefield, new methods have been incorporated in countless emergency rooms around the world, and countless lives have been saved.

Democratization of technology is also facilitated by the growing multinational nature of corporations. In terms of ownership and where corporate headquarters (and their subsidiaries) are located, and by virtue of the nature of the global supply chain, technologies will continue to spread, providing access to a growing percentage of the world's population.

Such democratization will not be without stresses and strains. In his 1999 book *The Lexus and the Olive Tree*, Thomas L. Friedman describes the tensions resulting from people's desire to cling to their roots and histories while also reaching for the technology represented by the Lexus car. Friedman's work served as a harbinger for what followed fifteen years later in the Arab Spring, which saw celebrations, protests, mass riots, and tensions across a wide swath of North Africa and into the Arabian Peninsula. Arguably, such a spark—which rapidly turned into a raging fire—could not have occurred without the social media platforms that served as the accelerant for this movement.

TECHNOLOGY DEVELOPMENT IS NOT WITHOUT RISK

History demonstrates that technology development is never without risk. Accidents do occur, and development does not proceed unimpeded, with one success leading to another and another until the technology is mature. Often, setbacks are a very real part of technology development. In fact, the history of technology development is that great risks have been undertaken in the name of scientific discovery and technological advancement.

Earlier, the development of powered human flight was discussed. Omitted from that discussion was the cost in human lives that contributed to the development of the aviation industry that humankind enjoys today. The price for pushing the envelope can be steep. Otto Lilienthal died in 1896 in Germany, and Percy Pilcher died in 1899 in Britain, both from glider accidents involving an inability to control their aircraft.[24] In fact, it was an article reporting on these lethal glider flights that attracted the Wright brothers to aviation in the first place.

The history of spaceflight demonstrates the dangers early pioneers faced in the quest for mastering this domain. Since the 1960s, more than twenty US astronauts have lost their lives in accidents, either in space or preparing for spaceflight.[25] And eighteen astronauts and cosmonauts from other parts of the world have died during spaceflight. The number of near misses is also noteworthy. Perhaps the best known of these incidents is the Apollo 13 mission, where a catastrophic explosion during the early stages of the mission required nearly rebuilding subsystems of the aircraft in flight to avoid loss of the crew. In a very real sense, the smooth accomplishment of the mission would have been nothing more than a footnote in history, while the redesign and construction of the control mechanisms and life-support systems during spaceflight surely goes down as one of the great technological feats in the history of humankind.

Edward Jenner's testing of his theory of smallpox on James Phillips placed the boy's life in great danger by exposing him to live smallpox virus. But his use of the scientific method in making observations, developing a hypothesis, testing his hypothesis, and developing conclusions was important to advancing the science of the day to prevent smallpox and promote vaccination in general.

The early pioneers of the Nuclear Age likewise took great risks exposing themselves to radiation and other hazards. Some of those risks are still being borne out today at Superfund sites, where contamination from radiological material and chemicals remains. In the early days at the dawn of the Nuclear Age, the exact risks of exposure to radiological material were not known, and the imminent danger faced by the country was judged to be worth the longer-term risks that were likely to ensue.

In the 1950s and 1960s, when the United States had an offensive biological warfare program, tests were done that included the use of simulants in lieu of the actual pathogens. The simulants were thought to be harmless, but years later, it was discovered that they led to increased risks to individuals and communities where the tests had been conducted. In some tests, volunteers were deliberately exposed to live biological pathogens to gauge their responses, better understand the progression of disease, and develop medical countermeasures that would be useful in protecting military personnel and civilian populations. During these experiments, several people were inadvertently exposed to highly virulent pathogens and succumbed to the diseases.

In fact, the US offensive biological warfare program essentially employed a structured R&D process that was essential for determining the relationships between the observed physiological effects of various pathogens on hosts, the best means of dissemination of the pathogens, and the impact of different climate conditions on the efficacy of the biological weapons. In one series of tests labeled Project 112, tests were done that varied the pathogens, the climatic conditions (e.g., arctic, temperate, tropical), deployment methods (e.g., sprayers, bombs, artillery), and formulations (e.g., wet versus dry) to determine the most effective combinations of biological weapons. The tests were conducted with urgency, given concerns about the ongoing NATO-Warsaw Pact standoff and the Soviet biological weapons program, and with only the best science of the day as a guide; these factors combined to introduce calculated risk into the US efforts.[26]

Today, as we contemplate these emerging technologies, we have a more complete understanding of risk, a more enlightened view of ethics, and greater sensitivity toward the rights of the individual. Studies such as the "Tuskegee Study of Untreated Syphilis in the Negro Male" for recording the natural history of syphilis to develop treatment programs are viewed as improper, unethical, illegal, and generally bad science.[27]

We see the same urgency in questions surrounding other technologies, such as unmanned aerial systems (or drones), artificial intelligence, social media, and autonomous systems. Drones had been on the market and in use by private citizens before policies, regulations, and laws were developed, leaving the Federal Aviation Administration to play catch-up in putting guidelines in place. Experts continue to define and contemplate a future with artificially intelligent systems. Revelations about the use and misuse of social media continue to be discovered, as limits on what is acceptable in terms of security of the data and individual privacy continue to be tested. What are the acceptable limits for use of these types of systems, and who should decide these limits?

Arguably, few technologies are under more scrutiny today than autonomous vehicles, which have been the subject of a great deal of discussion recently. Questions surrounding safety are taking priority, as they should during the early developmental period. We should expect these questions to continue until society has been made comfortable with the routine use of autonomous systems in areas that previously were the domain of humans. However, there will come a time when a mixed fleet of vehicles is on the road, consisting of human-controlled and autonomous vehicles. This is likely to be an extremely dangerous time as the human decision-making process and the autonomous vehicle's algorithms "collide" in a single transportation network with vastly different control mechanisms.

For the sake of scientific and technological discovery, risks have routinely been undertaken. Many resulted from an incomplete understanding of the natural phenomena on which the technology depended. Such was the case in the US offensive biological warfare program. In other cases, the risks were perhaps more calculated, as in the case of Edward Jenner exposing James Phillips to smallpox. Despite the outcome in this case, more enlightened modern concepts such as informed consent laws would make Jenner's actions illegal today. In other cases, our general interpretation of legal and ethical imperatives and risk tolerance have coalesced to make the Tuskegee study unthinkable today. Still, risks often continue to result from an inability of the policy, regulatory, and legal instruments to sense when limits should be considered and then implement such measures in necessary timelines.

Later, we will discuss methods that have been developed to account for costs, benefits, and risks of technology development to ensure that discovery results from proper science and technology that conforms to the highest practices or legal, ethical, and moral standards while minimizing the risks to individuals and society.

LEARNING TO READ THE WARNING SIGNS IS ESSENTIAL

Getting technology wrong can be catastrophic. Failure to understand the physical phenomena governing a technology can result in erroneous and potentially dangerous conclusions. Not understanding the operational context within which the technology will be employed can also lead to catastrophic outcomes. Finally, inadequate design specifications can result in technologies that lack the resiliency to operate outside of the narrow operational envelopes for which they have been designed and built. Indeed, history provides numerous examples of getting it wrong.

The Maginot Line in France prior to World War II was designed to be an impenetrable fortress protecting France from a potential invasion. It incorporated heavy fortifications, large artillery pieces, a series of obstacles, and living areas for vast numbers of French troops. It was designed to serve as a deterrent and, should deterrence fail, as an impassable defensive barrier. Despite its heavy construction and concept for employment, the French failed to account for a German scheme of maneuver—the blitzkrieg—that called for bypassing and overwhelming the fixed fortifications the French had developed.

The crash of the *Hindenburg* airship at Lakehurst, New Jersey, in 1937 represents another major technology failure. A hydrogen leak from the hull of the airship was likely ignited in the electrically charged atmosphere, and the airship was destroyed in the ensuing fire. Another theory involves the use of incendiary paint that ultimately caught fire and caused the accident. Regardless of the exact cause, technology failure or failures contributed to the loss of the *Hindenburg* airship and the demise of the airship as a means of mass commercial transportation.[28]

The sinking of the "unsinkable" *Titanic* provides another example of getting it wrong, resulting in a catastrophic outcome. What originally appeared to be a minor collision with an iceberg became catastrophic by the fracture of the steel hull, the failure of the rivets, and flaws in the watertight compartments. The failures were compounded by human errors; communications difficulties; and the provisioning of the ship, which only provided lifeboats for a fraction of the passengers and crew. An analysis indicating that the *Titanic* could remain afloat for two to three days following a collision with another ship proved to be in error, and the ship sank in a matter of hours despite its "watertight" compartments. The technology also failed to account for an austere operational environment that included low temperatures and high-impact loading, which contributed to the brittle fracturing of the steel hull.[29]

Kodak film provides another interesting case study of getting it wrong. The failure to anticipate the effect of the digital era on the use of traditional film had a catastrophic impact on the company. At its pinnacle, Kodak was a market leader in film for all applications from individual tourist cameras to sophisticated cameras used for reconnaissance by military forces. Despite their long history in the market, Kodak did not envision a day when traditional film would no longer be required and would be replaced by digital media. When this day came, cameras and film were fundamentally changed, as was Kodak's line of business and share of the photography market. While Kodak remains a viable company, their market share has been greatly reduced with respect to digital imaging for consumers' needs.

In some cases, design features have contributed to spectacular failures. The history of cascading electric grid blackouts, for example, represents a technology failure mechanism designed as part of grid resiliency. Grid technology was intended to allow for graceful degradation should anomalies be detected. Yet on several occasions, major blackouts have occurred as the electrical grid was attempting to respond to increasing electrical demands, to more efficiently balance the operating loads, or to heal the network after damage was sustained. One such example occurred on August 14, 2003, when fifty million people in the Midwest and Northeast as well as Ontario, Canada, lost power for up to four days. In assessing this outage, experts concluded that failures were caused by a small number of outages in generating plants and by trees touching power lines in neighborhoods linked to cascading power transmission. Once these failures began, they were impossible to halt. In fact, the grid was operating as intended and was working to balance loads across its customers. It was the electronic rebalancing of the system that led to a broader failure.[30]

A common element in these cases—from the Maginot Line to the electrical grid—was that developers failed to understand the relevant operational environments and design their technologies to have the structure and resilience to operate within them.

Another major pitfall to be avoided is the "valley of death" that frequently occurs in technology development. This is very much related to the question of connectedness versus disconnectedness addressed previously. If a technology does not have a willing customer on the other side of the developmental process, the chances that it will cross the "valley" and become something that serves a practical purpose diminish greatly.

The phrase "willing customer" is not just about the desire for the technology but includes such factors as technology feasibility, cost, and amount of training time required for developing the capacity to use the technology, as well as other considerations and impediments that could prevent the practical purpose from being achieved. Several examples serve to illustrate this concern.

The Department of Homeland Security's Customs and Border Protection wanted to use aerostat balloons as observation platforms to detect groups of people attempting to cross the southwestern US border. The Department of Defense had used many of these systems during operations in Iraq and Afghanistan and was willing to provide them to CBP at no cost. However, doing so would have meant a large annual bill for contractors to monitor the border, fly the systems, and maintain the aerostats. These resourcing constraints limited the numbers of systems CBP could afford to operate and thus the amount of border surveillance coverage that the aerostats could perform. It was not that CBP was not interested in the technology but rather that the resources were simply not available.

In the case of the army's future combat system program, the customer was very interested in the technology, but the TRLs were simply not mature enough to be integrated into an operational system. In fact, even almost twenty years after the program was canceled, many of the technologies remain in early TRL states and therefore not suitable for inclusion in a prototype system, much less for use in an operational environment.

Even where the customer has an expressed need and the resources are adequate, obstacles may appear. In the development of vaccines, heavy regulatory burdens for either human or animal vaccines translate to a lengthy period for demonstrating safety and efficacy. The normal developmental life cycle for medical countermeasures (including vaccines) is approximately a decade unless "emergency use authorization" can be obtained due to operational necessity.

CONCLUSIONS

In examining the foundations of technology development, we have been leading to an overarching theme. Whether thinking about the S-curve and the recursive nature of technology or considering the effects of dual-use, disruptive technologies and democratization of technology or contemplating connectedness versus disconnectedness, we have been leading to the concept of convergence of technologies. History serves as a reminder that as technologies are developed, they are combined and recombined in unpredictable, reinforcing ways.

In effect, the future portends the continuing convergence of technology such that the whole becomes far greater—perhaps even exponentially greater—than the sum of the parts. Through this transformative process, we come to see that technologies and combinations of technologies lose their original forms as they inherit other technologies. Consider the automobile that has become digitally enabled. While its lineage can be traced to the first automobiles, its utility individually and as part of the broader transportation system has transcended the original capability. The once-single-purpose conveyance—now augmented with communications, GPS, and specialized sensors—has become an extension of our lifestyles and, for some, a mobile office. With the infusion of autonomous systems, this transition will be complete.

Predicting the future of technology and even which technologies will be disruptive is an important question for those who specialize in scouting for innovative technologies. Such predictions could also be quite lucrative for those who predict correctly. However, technology forecasting is made more challenging by the nature of technology as assessing future technologies requires both complex nonlinear and combinatorial thinking. Getting this question right also serves to separate the innovators and early adopters from the rest of the pack.

In chapter 6, a methodology will be proposed for systematically examining technologies in the past and today, in the hopes of identifying the likely trajectories of technologies in the future. Looking forward, the foundations identified in this chapter can be very useful to these assessments. First and foremost, the structured way of thinking about technology in the form of the S-curve and Brian Arthur's evolutionary technology principles provide the bedrock for thinking about future technology development.

Second, risks are inherent in any research and development. Scientific discovery is not without dangers. The path to developing successful technologies is not linear and may result in many twists and turns before a technology can be put to practical use. Along the way, technologists would do well to consider the effects of the three D's (disruptive, democratized, and dual-use technologies). Disruptive technologies can make others—including well-entrenched incumbents—obsolete almost overnight. Democratization will continue to lead to the erosion of technological superiority and increase technology diffusion. All technologies are dual use and should be considered for both their societal benefits and their potential for malicious use.

Finally, the question of technology connectedness versus disconnectedness will continue to plague the science and technology communities. Natural tensions exist between those desiring to gain fundamental understanding of natural phenomena versus others from operational communities, who will seek to see science and technology applied toward practical uses. In this push and pull, the operator will likely have the upper hand, claiming operational necessity, which will be more compelling than the need for enhancing basic understanding. Still, sacrificing long-term benefit for the sake of operational necessity should not become the default priority for technology development.

CHAPTER 4

THE MODERN AGE OF TECHNOLOGY

Rehashing our earlier discussion of science and technology, research and development, and innovation and transformation would do little to increase our understanding of technology development. Rather, in this chapter our focus will be on understanding the evolution of the "modern" age of technology, which we will loosely define as beginning in the World War II period.

Early science and technology emanated from humankind's desire to solve problems and ensure survival. With basic needs met, humankind could turn to more complex questions, such as seeking to better understand the natural phenomena that govern our world. Learning by doing and rudimentary experimentation were replaced with a systematic scientific method, designed to provide scientific discovery with repeatable results.

Along this path, the World War II period was a dynamic time when fundamental changes in global security, including the rise of the United States as a world power, required new mechanisms for retaining and expanding this influence. One of these mechanisms was the establishment of a US and global scientific and R&D infrastructure.

VANNEVAR BUSH AND *SCIENCE, THE ENDLESS FRONTIER*

Vannevar Bush's report *Science, the Endless Frontier* at its core is a continuation of the connectedness versus disconnectedness debate. It was largely based on Bush's observations and his belief that a connected model for science and technology would result in a system in which most of the effort and resources would be dedicated to near-term problems rather than longer-term scientific questions and fundamental understandings. Said another way, the practical issues would not leave room for examining the more theoretical questions, those with a longer time horizon.

Bush's report was prepared in response to President Franklin D. Roosevelt's questions (fig. 4.1). It is noteworthy that a majority of the issues implied in Roosevelt's questions concerning the conduct of science and technology remain relevant today. For example, in the first question, Roosevelt contemplates how to share the benefits of technology. The sentiment is no different than that expressed in President Barack Obama's initiative that promoted sharing the benefits of government-sponsored research and development.[1]

Spending on R&D in the life sciences—as raised in Roosevelt's second question—continues to take a leading role in government-sponsored research and development. Today, in fact, government-sponsored research in the life sciences represents the largest share of non-DoD government spending on R&D.

Regarding the third question, on promoting research conducted by public and private organizations, these same topics are of interest today. Maintaining the balance between public and private organizations in science and technology remain an important topic. Programs such as Small Business Innovation Research (SBIR) and Broad Agency Announcements (BAA) are in use today by the government to inform industry about requirements for innovation and for research, development, and acquisition, respectively. More on both the SBIR and BAA programs will be discussed later.

The final question regarding recruitment of talent and workforce development continues to challenge government leaders today, just as in the post–World War II era. Today, major concerns about STEM education and recruiting and retaining a cybersecurity workforce highlight current significant issues.

Through these questions, Roosevelt was asking Bush to develop a plan for a postwar federal role in R&D. While some have speculated that Roosevelt's questions were drafted by Bush, they were the right ones in 1945, and they remain highly relevant to the scientific and technology development community today.

Figure 4.1. President Roosevelt's questions to Vannevar Bush regarding the future of science

(1) What can be done, consistent with military security, and with the prior approval of the military authorities, to make known to the world as soon as possible the contributions which have been made during our war effort to scientific knowledge?

(2) With particular reference to the war of science against disease, what can be done now to organize a program for continuing in the future the work which has been done in medicine and related sciences?

(3) What can the Government do now and in the future to aid research activities by public and private organizations?

(4) Can an effective program be proposed for discovering and developing scientific talent in American youth so that the continuing future of scientific research in this country may be assured on a level comparable to what has been done during the war?

Source: Franklin D. Roosevelt, to Dr. Vannevar Bush, November 17, 1944, Washington, DC.

Bush had come to believe that World War II would be won through the mastery of advanced technology. He further believed that such technology development would be more effectively conducted by academia and in civilian labs rather than through the traditional military acquisition channels. In 1940, Bush proposed this idea to President Roosevelt, who agreed and appointed him to head the National Defense Research Committee, which a year later was renamed the Office of Scientific Research and Development (OSRD).[2]

His position as head of the OSRD during World War II provided an important window into the development of all wartime military R&D.

Through these experiences, Bush concluded that in a connected model of R&D, US efforts in the basic sciences would suffer. Further, he concluded that doing so would jeopardize US standing more broadly. To this end, Bush stated, "A nation which depends upon others for its new basic scientific knowledge will be slow in its industrial progress and weak in its competitive position in world trade, regardless of its mechanical skill"[3]—thus indicating his belief in an inextricable link between scientific discovery and economic vitality.

The report came out in July 1945, shortly after Roosevelt's death. The report and the responses to the four questions were based on recommendations from the several committees established to consider the president's request. In the last sentence of the transmittal letter back to the president, Bush telegraphed his response by concluding, "Scientific progress is one essential key to our security as a nation, to our better health, to more jobs, to a higher standard of living, and to our cultural progress."[4] In short, Bush passionately called for retaining a strong role for government in R&D in the post–World War II period.

The report came out at a time when an important restructuring of the R&D performer community had just taken place. Up to that point, a majority of R&D was done by industry and related to purposeful efforts that would eventually result in the establishment of acquisition programs. In other words, the R&D was intended to mature technologies to allow for their inclusion in commercial applications, as well as military systems. By one account, corporations conducted 68 percent of the R&D, government 20 percent, and academia only 9 percent. However, during the war years from 1941–1945, government spending on R&D conducted in academic institutions accounted for almost 75 percent of all government R&D dollars and had climbed to $450 million per year. The result was a fundamentally restructured science and technology community that largely remains today and is responsible for government-sponsored R&D.[5]

In the *Science, the Endless Frontier* report, Bush also responded directly to President Roosevelt's four questions. For each of the areas covered, wartime successes were linked to science and technology developments.

His greatest enthusiasms were clearly regarding the need for continuing scientific progress, retaining a role for government in science and technology, and ensuring renewal of scientific talent. He also called for sharing the "vast amount of information relating to the application of science to particular problems," where he asserted that this information would be useful to industry. This program of action also called for the establishment of a new agency "adapted to supplementing the support of basic research and colleges, universities, and research institutes."[6]

Just as Bush's report was being released, Congress was debating the future of national R&D institutions. The result was compromise legislation that would allow the OSRD charter to expire on December 31, 1947, and be replaced by the National Science Foundation (NSF). President Truman vetoed the bill because it did not give authority to the president to name the director of the NSF.[7] After three years of debate, legislation establishing the NSF was finally passed in 1950.

During the intervening period between when the OSRD sunsetted and the NSF was established, the Office of Naval Research under the US Navy largely assumed the role the OSRD had been filling. ONR's chief scientist was nominated and confirmed to be the first director of the NSF in March 1951.

Over the twelve-year period from 1941–1953, the science and technology establishment in the United States had been transformed. The ushering in of the nuclear era and signing of the National Security Act of 1947 resulted in a growing military-industrial complex that required significant R&D investment and capabilities. Corporate funding diminished as a percent of overall R&D, with the federal government now accounting for approximately 60 percent of R&D funding. Academia became a major recipient of the R&D funding coming from government sources. As an example, by 1953, the Department of Defense and the Atomic Energy Commission (AEC) were spending 98 percent of all the money being spent nationally on physics research.[8]

Another part of this transformation was the massive footprint created as part of the Manhattan Project. These facilities would largely become the Department of Energy laboratory system. Today, the Department of Energy laboratory complex includes seventeen laboratories that are maintained as government-owned and contractor-operated facilities. Sixteen of these labs have been established as federally funded research and development centers (FFRDC), meaning they occupy a space between federal employees and contractors. As FFRDCs, the labs have special relationships with the government and are required to conduct R&D in areas where there are needs that cannot be met by either the government or the private sector.[9] These laboratories have unique capabilities (requiring significant capital expenditures), including photon light sources, high-flux neutron sources, nanoscale science centers, high-performance computing facilities, high-energy physics and nuclear physics facilities, biological and environmental facilities, and fusion research facilities, just to name a few. Several also operate their own nuclear reactors and have facilities for conducting high-explosive tests and simulated attacks on industrial control systems, as well as for 3-D fabrication and production design.[10]

THE RISE OF THE ARPAS

ARPA stands for Advanced Research Projects Agency. Secretary of Defense Neil McElroy established ARPA in 1958. The name would later be changed to the Defense Advanced Research Projects Agency, or DARPA, as it is commonly called. To truly understand the importance of the establishment of DARPA, one needs to go back to post–World War II history.

In World War II, the United States and our allies had prevailed over the Nazi and fascist threats. Science and technology, mass production of enormous quantities of matériel to outfit newly created units and replace destroyed equipment, and the ability to project power globally contributed greatly to the victory. Through the establishment of the National Security Act of 1947, America proclaimed its readiness to be a global leader.

Alliances between the Western European allies and Soviet Union began to unravel immediately following the war. The post–World War II resurgence of communism and historic animosities established a new series of challenges and threats to be confronted. The Soviet Union, China, North Korea, and communist insurgencies and movements led to instability throughout Europe and Asia. To thwart the spread of communism, President Truman announced the Truman Doctrine in 1947, which established that the United States would "provide political, military and economic assistance to all democratic nations under threat from external or internal authoritarian forces."[11]

The Soviet launch of the first artificial satellite, the Sputnik-1, in 1957 came as a shock to US leaders and citizens alike. Moreover, it served as a wake-up call regarding the advances the Soviet Union had made in technology development. The launch of Sputnik also intensified an already-tense arms race and increased Cold War tensions, leading to a growing perception of a "missile gap."[12] In fact, some have credited Sputnik with being the catalyst for Kennedy's "man on the moon" speech before Congress and the ensuing race to the moon.[13]

One account describes Sputnik's influence as follows:

> There was a sudden crisis of confidence in American technology, values, politics, and the military. Science, technology, and engineering were totally reworked and massively funded in the shadow of Sputnik. The Russian satellite essentially forced the United States to place a new national priority on research science, which led to the development of microelectronics—the technology used in today's laptop, personal, and handheld computers. Many essential technologies of modern life, including the Internet, owe their early development to the accelerated pace of applied research triggered by Sputnik.[14]

With Sputnik-1 as a backdrop, President Dwight D. Eisenhower made several subtle and not-so-subtle changes to the US R&D establishment.[15] He established the first ever Special Assistant to the President for Science and Technology in 1957 as a direct report to the president. The National Aeronautics and Space Administration was founded in 1958. Funding for the National Science Foundation was transformed and the budget tripled to \$159 million by 1959. President Eisenhower also placed increased emphasis on the teaching of science and mathematics in American school curricula and strongly supported the National Defense Education Act of 1958, which included these provisions.[16] And President Eisenhower established DARPA on February 7, 1958, with the mandate to "avoid technological surprise."[17]

The establishment of DARPA saw the consolidation of the army's and navy's R&D programs under this newly formed organization. Inherent in DARPA's mission and operating style was to do things differently and not accept the status quo, looking for opportunities to meet future challenges as well as to create new futures. DARPA was granted special authority in resourcing—for both funding and access to personnel—to enable this freewheeling approach. And for more than sixty years, DARPA has operated under this special authority that provides its leadership and technologists the freedom to pursue high-risk, high-payoff programs. In short, DARPA embraces a culture that promotes taking chances in pursuit of its "singular and enduring mission: to make pivotal investments in breakthrough technologies for national security."[18]

DARPA's mission requires the pursuit of "transformational change instead of incremental advances." With a small staff of around 220 people and a flat organizational hierarchy consisting of six technical offices, the workforce pursues innovative concepts that can be turned into practical capabilities. DARPA's focus is on basic and applied research and technology development considered imperative for helping technology make the transition from the laboratory to the field.

A key to DARPA's success has been identifying partners across the broad range of R&D organizations. Those successes have included game-changing military capabilities such as precision weapons and stealth technology, as well as advances that have benefited both the military and civilian society, such as the internet, automated voice recognition and language translation, and Global Positioning System receivers.[19]

Both by design and evolution, DARPA represents a connected model of conducting science and technology. While many of their technologies began for the purpose of improving military capabilities, DARPA has also had success in developing dual-use capabilities that clearly have great benefits for civilian applications and humankind in general.

To promote the connected R&D model and provide agency officials a foundation from which to think through and evaluate proposed research programs during his tenure, George H. Heilmeier, DARPA director from 1975–1977, developed a set of questions to evaluate proposed research programs. The list became known as the Heilmeier Catechism:

- What are you trying to do? Articulate your objectives using absolutely no jargon.
- How is it done today, and what are the limits of current practice?
- What is new in your approach and why do you think it will be successful?
- Who cares? If you are successful, what difference will it make?
- What are the risks?
- How much will it cost?
- How long will it take?
- What are the mid-term and final "exams" to check for success?[20]

Similar evaluation criteria have been developed for portfolio reviews to assess R&D portfolios. The questions are designed to force the program managers to examine the goals of the potential project, assess what is likely to be accomplished through the study effort, and evaluate whether it is likely to have operational significance.

While DARPA was one of the first post–World War II R&D organizations, over the course of the last sixty years, the R&D enterprise for defense and nondefense applications has grown considerably. The Department of Defense lists a total of seventy-six labs: two DoD, thirty-six army, twenty-four navy, and fourteen air force facilities.[21] The work done in these facilities covers a wide range of activities, including basic and applied research and early technology development, as well as supporting technology development for risk reduction in DoD acquisition programs. The military also depends on FFRDCs, including the DoE laboratories and University Affiliated Research Centers, in addition to collaboration with international partners and industry.

By way of an example, the Air Force Research Laboratory (AFRL) has approximately 3,500 scientists and engineers, of whom 2,800 are civilians and 650 are military. AFRL has a total operating budget of $4.8 billion, of which $2.5 billion comes from the air force and the remaining $2.3 billion from other sponsors in the US government such as DoE and DARPA. Its mission is to lead the "discovery, development, and integration of warfighting technologies for our air, space and cyberspace forces."[22] Each of the services has similar organizations for satisfying their technology development requirements. In addition to these DoD labs, these internal capabilities are augmented through partnering to satisfy the full range of the department's R&D requirements.

The other departments and agencies have developed their own versions of DARPA. While they do not all function in the exact same manner or with the same goals as DARPA, they play a critical role in bringing S&T, R&D, and I&T to their respective organizations. To this end, their mandates come from their individual departments and agencies.

The DoE has the Advanced Research Projects Agency-Energy (ARPA-E), established in 2009, which pursues a research portfolio of both programs for developing near-term capability for current systems and projects that look longer-term for developing higher-impact technologies. ARPA-E focuses its efforts on energy applications, seeking to develop interagency partners and start-up technology companies with an interest in the energy sector. By design, ARPA-E works in areas where other offices and energy partners are not working. In 2017, its budget was $306 million.[23] Since 2009, ARPA-E has received more than $1.5 billion in funding and lists having funded 580 potentially transformational energy technology projects.[24]

The Intelligence Advanced Research Projects Activity (IARPA) supports the Intelligence Community (IC) with research and development. IARPA conducts both classified and unclassified research and works for the Director of National Intelligence.[25] In conducting scientific discovery and technological advancement, IARPA also emphasizes not attempting to compete with other offices and partners within the IC.

It is noteworthy that each of the other sixteen members of the IC also retains internal capabilities for conducting R&D. The budgets for the IC R&D remain classified.

The Department of Homeland Security has a Homeland Security Advanced Research Projects Agency (HSARPA) designed to conduct a broad range of R&D activities supporting the Homeland Security Enterprise (HSE).[26] HSARPA is in the department's Science & Technology Directorate and focuses on identifying, developing, and transitioning technologies and capabilities to counter chemical, biological, explosive, and cyber threats, as well as protecting the nation's borders and infrastructure.[27] The funding for the entire S&T Directorate was approximately $627 million in 2018.[28]

For the Department of Health and Human Services (HHS), the National Institutes of Health (NIH) fulfills the ARPA role. As highlighted previously, government funding for the life sciences is higher than for any of the other science and technology fields. NIH has a budget of $32.3 billion for medical research. Most of this "funding is awarded through 50,000 competitive grants to more than 300,000 researchers at more than 2,500 universities, medical schools, and other research institutions in every state and around the world."[29]

Another important office in the realm of science and technology is the Office of Science and Technology Policy (OSTP), which was established in 1976 and is part of the Executive Office of the President. The director of the OSTP advises the president on science and technology matters. Additionally, the director traditionally chairs the President's Council of Advisors on Science and Technology (PCAST), which has a broad mandate for providing advice on S&T issues, and the National Science and Technology Council (NSTC), which coordinates the Federal R&D enterprise.

Throughout the government, offices responsible for managing, monitoring, or overseeing R&D efforts have been established. For example, in the DoD, each of the services has a science advisor and a separate organization responsible for acquisition, technology, and logistics. The DoD spends approximately $70 billion annually for R&D, which is approximately half of the government's R&D spending.

No discussion of R&D in the United States would be complete without highlighting the National Science Foundation's current role in R&D in the United States. From its roots in 1950, the NSF continues "to promote the progress of science; to advance the national health, prosperity, and welfare; to secure the national defense."[30] With an annual budget of $7.5 billion in 2017, the NSF provides funding for 24 percent of the federally supported basic research conducted by America's colleges and universities. The NSF also plays a role in tracking research in the United States and across the globe. The NSF promotes research through a competitive grant process, awarding approximately eleven thousand of the forty thousand proposals received annually.[31]

While it might be tempting to call this conglomeration of US R&D organizations an enterprise, that would be overstating the degree to which they coordinate and collaborate as they conduct scientific research and technology development. Certainly, some of these organizations work together on specific projects, but there is also a spirit of competition between them. Even in the DoE labs or within the R&D centers of the military services, full transparency is not common. Furthermore, the smaller, unaffiliated R&D organizations in operation could hardly be characterized as being part of an enterprise, as much of their work remains unknown to the other R&D organizations unless a breakthrough occurs.

Given both the funding for and the performers of the R&D, one can observe that these organizations represent only a small fraction of the national R&D that is conducted. As highlighted earlier, the combination of centralized government spending and decentralized R&D across the United States in industry, academia, and even by private citizens makes placing a single figure on national R&D a virtually impossible task. These funding methods also provide challenges in attempting to focus national R&D efforts.

IN SEARCH OF INNOVATION

In our earlier chapter on the technology lexicon, the point was made that there are differences between science, technology, research, development, innovation and transformation. Yet, as also highlighted, identifying the boundaries and distinguishing between these activities is not so clear. In fact, if one were to examine government funding for R&D, it would undoubtedly include spending on innovation activities such as experimentation and technology demonstrations.

Earlier, we defined innovation as a process for generating, promoting, and operationalizing ideas, requiring a culture that promotes taking prudent risks and having the freedom to pursue alternative solutions. Today, innovation is frequently being used to offset requirements for conducting scientific discovery or technology development. By placing small, operator-focused organizations in hotbeds of technology development, the federal government hopes to be able to capitalize on the discovery and advancement that are being or have already been done.

This approach seeks to develop an interface with individuals and organizations conducting S&T and/or R&D in order to identify early-stage technologies that can be adapted, technologies requiring only small tweaks that could be incorporated into operations, and even full systems that could be integrated "as is" to solve operational problems. If one thinks back to the S-curve discussion, in a sense, the federal government is trying to start farther up the curve rather than having to invest in early research and development.

The need for such technology adoption and absorption is related to the fundamental shifts in the funding and conduct of R&D. Whereas the federal government previously was the largest funder of R&D, today, industry has surpassed the federal government and is by far the largest source of funding. Furthermore, the rate of change of technology makes it infeasible to conduct basic and applied research programs in all areas where the government has an interest. The implication is that the federal government must become a better consumer (rather than driver) of technology to remain on the cutting edge.

The Revolution in Military Affairs discussed previously was an effort to bring innovation into military operations. The earlier efforts from the 1990s led to the growth of new organizations and additional activity across key areas of interest. More recently, Secretary of Defense Ash Carter established the Defense Innovation Initiative (DII). The goal was to establish innovative ways to "identify the 'art of the possible' for national security."[32]

The Defense Innovation Unit Experimental (DIUx) was established in 2015 to focus on five areas: artificial intelligence, autonomy, human systems, information technology, and space. Consisting of approximately fifty people, DIUx claims to have "launched and sold companies backed by tier-1 VCs [venture capital companies]; led teams at the Joint Staff, the Office of the Secretary of Defense, and the White House; served with our military around the world; and helped build some of Silicon Valley's most iconic companies."[33] In 2018, DIUx was changed to DIU, removing "Experimental" from the program in recognition of the perceived success of the concept and the need to make this a standing, more permanent organization.

More recently, DoD has created the MD5 National Security Technology Accelerator, which is a public-private partnership between DoD, New York University, and a broader network of top US research universities. This technology accelerator is designed to bring military and civilian first responders together to develop solutions for humanitarian assistance and disaster relief challenges in urban environments. The goal is to promote a "network of innovators and entrepreneurs with the incentives, expertise, know-how and resources required to successfully develop, commercialize and apply DoD technology."[34]

The quest for innovation is not a DoD priority alone. The Intelligence Community created a venture-capital company in 1999 called In-Q-Tel for attracting and promoting leading-edge technologies. It is a public-private partnership with the goal of identifying interesting technologies, providing an environment for maturing those technologies, and supporting their transition to use. IQT has grown both as a concept and with the partners it attracts.

The chief technology officer at Health and Human Services recently announced the establishment of an HHS IDEA Lab to "encourage and enable innovation" within the department. The services provided by the IDEA Lab include an excite accelerator, a ventures fund, the HHS Open Innovation Program, and entrepreneurs-in-residence.[35]

The quest for innovation is not unique to government either. A source examining business innovation highlights that requirements for innovation were growing beyond its roots in R&D: "Strategic planning, business units, think tank/incubator, IT and the innovation committee all were listed ahead of research and development as places where innovation is spotted and nurtured in companies."[36]

Formal and informal innovation hubs continue to be established. DoD's DIUx was one such incarnation, but others come from places such as Menlo Park, where small, innovative start-ups freely mix with larger, more established companies—one looking to develop a new concept or technology, the other looking to acquire a "component" that can be integrated into one of their products. Others are regionally based and growing to meet an emerging demand in places one would expect (and others perhaps not so much): United States, Saudi Arabia, Europe, China, and Africa, to name a few. One report describes the innovation hub opportunities in Africa as "offering learning, networking and collaboration opportunities as well as work spaces for individuals and small start-ups."[37]

To demonstrate the fluidity and ever-changing nature of the quest for innovation and the perceived leadership in many technical areas, during the Clinton administration, a collection of programs through the Defense Conversion, Reinvestment, and Transition Assistance Act of 1992 was established to provide an avenue for companies focused on delivering defense-related technologies to "channel their research into commercial products and to improve the productivity of US companies."[38] In an odd twist, the goal is now opposite—to find mechanisms for the military to take advantage of technologies initially developed for civilian applications. The nonmilitary departments and agencies are also pursuing these same types of outcomes.

Finally, while no single accepted definition of *innovation hub* has been provided, one source questions, "Is it [tech innovation] just a trend term replacing 'incubators,' 'R&D labs,' 'science parks,' 'technopoles,' or 'training facilities'?"[39] So while the definition might be unclear, what is clear is that a global movement is ongoing in which governments and individuals are seeking greater access to S&T, R&D, and I&T and that innovation hubs are serving as one of the vehicles for this journey.

In today's ever increasingly technologically enabled world, we are witnessing the purposeful growth of a kind of sophisticated neural network connecting operators, technologists, developers, and scientists. Those both willing and able to develop these networks will likely be best positioned to take advantage of the opportunities this unique environment offers.

WHO IS ACTUALLY DOING THE R&D?

When I was in the DHS S&T Directorate, the senior people in the department's operational organizations[40] would often ask for technical assistance in satisfying real-world problems. Perhaps it was assistance in developing a new process for improving operational throughput, or an elusive big data problem, or a question about development of a technology for improving situational awareness along the southern border of the United States. Each of these problems had very different operational requirements and desired outcomes. Some were looking for process improvements; others for software to apply to their large data sets; and still others needed hardware, software, and systems to be integrated into a capability for satisfying the needs of their complex operational environment.

Invariably, the question would come up: who was the best performer to assist with satisfying these operational needs? The answer: it depended. Each performer had different skill sets, business models, timelines for development, and costs. The most appropriate performers for a task were as varied as the operational requirements where some components required assistance with applied research issues, others with technology development, and still others required advice and support for obtaining commercially available products.

One such example pertained to big data issues to support the volumes of information generated and used daily in DHS. It had become clear the department had numerous issues requiring data solutions, from screening and vetting of passenger manifests to targeting cargo ships and containers for inspection. To better understand the range of requirements and begin to support the customer, we conducted a survey that requested each operational component submit a list of their big data requirements. The survey was simple, requiring no more than identifying the component requiring support, a brief description of the problem, and a point of contact. The results of the survey were enlightening.

Across the department, forty-four big data problems were submitted for consideration. When we evaluated the submissions, we discovered that the requirements could be binned into four categories: those for which commercial solutions were already available and could be purchased off the shelf, those for which there was a similar commercial system that could be modified to fulfill the operational requirements, those considered tough problems requiring R&D, and finally those that were difficult to understand and required additional interface with the customer to better appreciate the problem. Interestingly, each of the four categories accounted for approximately 25 percent of the submissions.

This experience reinforced that improvements and building capabilities in organizations do not necessarily require R&D. In some cases, it comes down to having the right neural networks in place linking operators, technologists, developers, and scientists.

Elaborating on all R&D organizations within the United States is certainly not possible. Nor would it get us any closer to understanding trends in R&D. However, it is useful to identify the key organizations within the R&D community and understand their respective roles. This knowledge becomes very useful when looking to assist customers with their R&D requirements for fulfilling operational needs.

Before directly answering this question, a bit of historical perspective is in order. Early manufacturing companies conducted most of their R&D internally. Say a company decided to manufacture a new tool. The charge for developing, testing, and producing the tool was the responsibility of a single line of business within the company. All work was done internally, and when the tool was ready, it was taken to market, where it was sold. This trend continued even once industry began to develop more formal technology development processes.

An example can be found in the history of Bell Labs, which in the mid-1920s became the technology development component of American Telephone and Telegraph (AT&T). Initially, what could be considered early R&D was all done in-house by Bell Labs scientists and engineers. These same scientists and engineers began their introduction to AT&T (and the lab) by learning how to climb telephone poles, string wire, and answer switchboards. Only after mastering these skills would they begin working on more technical areas, such as perfecting signal amplification and developing the vacuum tube. They were also the group responsible for the development of the individual components (or technologies) of the vacuum tube, such as the cathode, grid, and anode.

Early descriptions of Bell Labs R&D showed technology advancements based more on trial and error than scientific method. In developing methods for signal propagation, a variety of materials and configurations were attempted before finding combinations that supported signal dissemination over long distances.[41] It would have been difficult to delineate the science and technology, research and development, and innovation and transformation in this technology development effort. The work was all purposeful, with a use case clearly in mind. The Bell Lab employees were educated as scientists and engineers but were indoctrinated through their introduction to the company with an understanding of the operational problems.

This method of in-house R&D largely continued to work through World War II. However, as fields of technology grew and the complexity of the work increased, it became necessary to find specialists who could work in technical areas that may not be resident within the company.

In referring to the complexity of systems, think how technologies have evolved from just hardware to hardware and software and now to hardware, software, and embedded communications. This represents more complexity by orders of magnitude, and it begins to explain why R&D has become more specialized. To respond to this increasing complexity of technology, companies responded by changing their approaches to conducting R&D. One can think of these changes in terms of four generations.

Generation 1.0: internal company assets used for all R&D

Generation 2.0: increasing complexity requiring use of specialists for certain technologies

Generation 3.0: conscious management decision to focus on core competencies and outsource for specialties to save money

Generation 4.0: decision to use strategic technology sourcing for understanding of and access to complex and specialized technologies that can be brought forward for use at lower cost than internal development

Even large integrators have been looking for opportunities to outsource either technical components or entire areas of R&D that are not part of their core competencies. Another strategy some have employed has been to acquire R&D capacity through mergers and acquisitions. This can be particularly attractive if the company being bought has access to the intellectual property that a large integrator might require.

It is interesting to note that the military is following largely the same generations with respect to providing R&D for the force. The DIU program could be considered a Generation 4.0 application for the military. This characterization of generations is not to imply that corporations or the military no longer do internal R&D but, rather, that they are more selective in the early research and development, relying more on outsourcing and specialization. Such an approach is particularly attractive in areas where corporate R&D leads the military, such as in information technologies, big data solutions and open-source (including social media) applications.

If your operational problem requires R&D and therefore cannot be procured commercially, selecting the best performer requires understanding of the skills, costs, and benefits of each. However, there are no absolutes in making judgments about which performer will be best suited for the type of R&D required. Undoubtedly, exceptions will surface as one begins to become more familiar with R&D performers within your neural networks. Additionally, many of the same types of R&D performers that reside in the United States are also available internationally.

The first category are FFRDCs and internal labs within departments and agencies. Examples of these organizations include the DoE national security labs, labs within the military services, and specialized DHS labs operated by the S&T Directorate such as the Transportation Security Laboratory and National Biodefense Analysis and Countermeasures Center (NBACC).[42]

These organizations are best at multidisciplinary, complex R&D problems. Their capabilities generally involve basic and applied research and development activities that include preacquisition support. They normally have innovative capabilities that are beyond the scope and depth of academia and industry. Many of these FFRDCs include large footprints, with a variety of sophisticated laboratory and testing equipment. The customers for these FFRDCs are normally government departments and agencies, although in the case of the DoE labs, they do have a Work for Others program that allows for support to nongovernmental entities as well. FFRDCs are prohibited from competing with the private sector. Given this description of an FFRDC, it will likely come as no surprise to the reader that developmental capabilities in high-performance computing (HPC) at two of the DoE labs— Lawrence Livermore and Oak Ridge National Labs—have resulted in the development of two of the three largest HPC systems in the world; a facility in China is the other lab in the top three, and their positioning within the top three varies as new technologies and modernization of their HPC capabilities evolve.

The next category is academia. Colleges and universities provide a ready source of basic and applied research and early-stage development activities across a wide range of disciplines.[43] They are also important sources of ready intellect to be applied toward problems and should be considered a low-cost source of high-quality talent.

However, in contrast to the FFRDCs, they generally have smaller facilities and less state-of-the-art equipment. R&D requiring bench science, algorithm development, and research to gain a better understanding of natural phenomena is well within their sweet spot, while preacquisition support for major military system acquisitions would likely be beyond their capabilities. Of course, having one of these universities examine a small piece of a major acquisition program might also be well within their core competencies.

In the Department of Homeland Security, we had at any one time ten or so academic centers of excellence (COE) that could touch more than two hundred universities engaged in R&D in support of department requirements. Some of the programs undertaken by these COEs included development of an algorithm for detecting homemade explosives, research on various foreign animal diseases, physical screening techniques for large venues such as stadiums, optimization of patrolling using operations research techniques, and best practices for border security solutions.

In discussing industries, it is important to differentiate between the size and types of companies. In terms of size, we will discuss large integrators, medium-sized companies, and small businesses. The types of companies will be further differentiated by those that have a requirement for an R&D function (i.e., those developing hardware and software products) and service companies, which are normally sources of manpower. We will not discuss service companies.

Large integrator firms are likely very familiar to the reader, as they include such corporations as Northrop Grumman, Boeing, Lockheed, Siemens, and General Electric, to name a few. More recently, they have been joined by the likes of Google and Amazon, which have branched out from their original roots into areas such as the development of autonomous vehicles, grocery delivery, and cloud computing.

These large integrators are leaders in industries such as aerospace, defense, cybersecurity, health, intelligence, and logistics that require mastery of a broad range of disciplines. They also generally have sales in excess of $1 billion per year, and the integrator status reflects the wide-ranging expertise of their staff. While they certainly have the capacity for conducting all stages of R&D, their preference is often to rely on others for basic research, preferring to focus on applied science and developmental activities associated with preacquisition activities and formal acquisition programs. To be sure, they are profit-driven organizations with low tolerance for risk due to their fiduciary responsibilities to shareholders. They are also likely to be less flexible and agile due to these fiduciary responsibilities.

Today, one should not think of these large integrators as conducting all their basic research the way Bell Labs did back in the autumn of 1947 when they invented the transistor. This is not to say that they lack the capability to develop the equivalent of the modern transistor but, rather, that there is an opportunity cost associated with conducting such basic research. Their resources can be better spent, and shareholders will gain greater satisfaction by allowing the basic research to be done by small businesses with low overhead, high risk tolerance, and an innovative spirit. After the basic research has been completed, the intellectual property rights can be purchased, and in the end, the integrator has saved on the development of the technology.

Boeing Corporation is an example of a large integrator that still does R&D in its core competencies but also relies on bringing in companies as subcontractors or through acquisitions for other areas. Their internal Phantom Works and Horizon X are charged with looking toward new technologies and innovations in key areas, but they also look to outside entities for areas such as network security, where others likely have greater capabilities.[44]

When describing Boeing's goals for technology development, one senior research supervisor referred to their program as "tapping the trillion," meaning to take advantage of the broader technology community where it makes sense to do so. For example, Boeing works with specialized vendors for acquiring and developing its batteries, with the goal of getting better battery densities for systems, which would result in higher capacity, longer times between charges, and longer usable lives.[45]

In deciding on R&D investments with the normal budgetary constraints, Boeing divides its efforts into "must fund," where they have a competitive advantage and will conduct the R&D, and "relationship development," where the technology could be obtained from government labs, universities, or consortia. Concerning R&D, Boeing conducts "almost no" basic research and very little applied research (i.e., less than 5 percent of the total R&D budget) at the TRL 2 and 3 levels, and the remainder of the R&D budget goes to technology development for maturing technologies from TRL 4 to 5 or 6 before handing them off to the line business units for further development and incorporation into the product lines. The Phantom Works could become involved at TRL 4 for some technologies.[46]

In considering the technology life cycle, Boeing employs the three-horizon framework. The first horizon is the current airplane, looking at incorporating technologies that can extend the life of the aircraft, modernize the platform to provide increased capability, and improve performance. The second horizon looks at the design of the next aircraft to identify the technologies, systems, and designs that will be employed in the replacement of the current aircraft. The third horizon looks to the future to predict and understand what technologies and systems are likely to be available well beyond the next aircraft.[47]

In the next category, we will include medium-sized companies and small businesses. Medium-sized companies generally have annual revenue of between $50 million and $1 billion and are likely to be privately owned. The medium-sized companies have an eye toward growing their businesses and therefore will likely have greater risk aversion than a small company or start-up, but less than a large integrator. At the lower end of the annual revenue spectrum, these companies are likely to have issues with manufacturing large quantities of their product line. Of course, the manufacturer's capacity could vary greatly depending on the product in question.

Small businesses have revenues of less than $50 million annually, are normally privately owned, have a higher risk tolerance, and are more likely to engage in basic and applied research. Small technology firms normally are challenged by getting beyond TRL 5 and therefore depend on strategic partners to get their product to market. For these companies getting their technologies to the higher TRL and moving their product to market, they are normally limited in their ability to deliver much more than prototypes or small quantities. This category also includes DIY entrepreneurs, who generally have limited funding and staff. The early Apple company launched by Steve Jobs and Steve Wozniak fell into this category, as the company famously started in a garage in Los Altos, California.

Recently, another type of industrial model is emerging that appears to be transforming R&D and more specifically the space industry. One such example is Elon Musk's SpaceX. Through his wealth, scientific and technology interests, and entrepreneurial spirit, he is challenging traditional methods of R&D and innovation. From PayPal to SpaceX to Tesla, he is collapsing the traditional timelines for R&D. One account describes this revolution this way: "Musk turned space exploration from a government monopoly into a believable, private endeavor, and turned electric cars from a purely environmental initiative into a sleek, fun product."[48] Musk's effect on the space launch industry most certainly must be considered disruptive.

Musk's full effect on the broader R&D field has yet to be determined. Whether others can replicate the mix of technical and entrepreneurial capabilities Musk brings to an R&D problem remains a question. Still, the SpaceX Falcon Heavy was the most powerful US rocket launched since the Saturn V, which took astronauts to the moon in the 1960s and 1970s.[49] In a further demonstration of how Musk's SpaceX activity is changing the market, the rocket is demonstrating capabilities such as reusable boosters, which will lower the cost of space launches, and rockets with the capability to support his plans for Mars colonization. SpaceX has already demonstrated lower costs for greater launch capacity. The Falcon Heavy launch cost starts at $90 million, and the smaller Falcon 9 starts at $62 million. Another competitor in this new commercial space market, the United Launch Alliance, has a smaller Atlas V, which is advertised at $109 million.[50] Still other space launch companies are emerging, looking to develop the capability to rapidly and cheaply launch constellations of CubeSats for a variety of applications. In short, a global competition for access to space is underway, led by this new R&D model. Other fields such as biotechnology, artificial intelligence and cybersecurity are seeing the same emerging trends.

Discussion of R&D in industry would not be complete without discussing two other types of organizations. The first is the investment community composed of venture capitalists, angel investors, and strategic partners. The investment community provides funding for small businesses and entrepreneurs. It also can be important for identifying cutting-edge technologies and tracking investments to calculate rates of return for the R&D that has been conducted. In-Q-Tel, discussed previously, is such a venture capital firm, with a focus on the US Intelligence Community. These firms can be a source of funding for small start-ups and are willing to take a longer view of their investment, but they will normally look for a payoff—either in the form of a high rate of return or partial ownership—should the R&D be successful.

The second are industry organizations, which include trade associations, regional innovation clusters, and chambers of commerce. The Institute of Electrical and Electronics Engineers (IEEE) is an example of an industry organization. These organizations play significant roles in establishing standards and representing the industry in question. They are uniquely positioned to understand trends, develop road maps, and establish policies that promote the health of the industry. They are also essential in guiding the development of R&D. While they do not develop technologies, they protect and promote the interests of businesses and key technology sectors. Consider how the establishment of an IEEE standard for electrical components has led to the growth of this industry. Without such standards, there would be no compatibility for electrical applications.

While the various R&D organizations have been described, this still really does not answer the question of where to go for R&D. The answer is likely to be related to the type of technology required, number of systems to be procured, and the state of technology readiness. If one is looking for innovation, it is far more likely to come from startups or small businesses. R&D for large, complex systems and those for which significant acquisitions are planned suggest that a large integrator will be required. That does not mean that start-ups and small businesses could not be used in the early R&D or for individual component technologies, but it suggests that at some point, a large integrator will likely become a necessity to bring the larger technology successfully to market.

THE EFFECT OF THE DEMOCRATIZATION OF TECHNOLOGY

Previously, we established that technology democratization was occurring throughout the world, providing greater access to technology to nations and individuals alike. While the United States continues to spend the most on R&D, others are beginning to catch up. Since 2002, China has seen near-exponential growth in spending on R&D and is within $80 billion annually of what the United States spends on R&D. Numerous countries are spending more on R&D as a percent of GDP than the United States. In fact, today, North America accounts for only 27.9 percent of all R&D spending.[51]

The point is not to complain about R&D spending totals or press for greater spending. Rather, it is to remind the reader that the United States is no longer in the preeminent position with respect to global R&D. While US R&D spending in real terms is still the highest of any nation, the gap has been closing quickly. We should expect these trends to continue as democratization continues. Other nations have observed the United States, gaining an appreciation for the relationship—as Vannevar Bush suggested in *Science, the Endless Frontier*—between national power and science and technology.

However, given this democratization of technology and the inevitable diffusion of technological capability, the United States should expect that leadership in certain areas of scientific study will likely erode and even be held by other nations in the future.

The question, then, is what to do about this shifting landscape. Here are some thoughts on managing this trend.

First, the United States must become a better consumer of technology and get used to living in a world where it might not lead in all technology areas. This will likely be necessary, as the cost for leading in all critical areas (and the subcomponent technologies) would undoubtedly be unaffordable. We have already seen this occur in several areas, such as biotechnology, where China leads in high-throughput sequencing capacity and the United States (at least for now) leads in synthetic biology. The same can be said of cybersecurity, where the United States, Russia, China, and Israel all have strengths in some sub-areas, but there does not appear to be one clear leader.

Second, setting priorities for science and technology will be imperative to ensure the highest priorities are funded, gain an understanding of where the gaps in R&D are, and assess where the work of others can be incorporated into the efforts of US government and industry. For areas where the United States wants to gain or retain a strategic advantage, infusion of resources, leadership focus, and developing priorities will be essential. However, setting priorities such as the publication of national strategies in technology areas such as cybersecurity and artificial intelligence will do little if these strategies are not appropriately resourced. Unfortunately, many White House or department and agency strategies often contain only objectives, with little detail on how they will actually be accomplished. In military terms, it is fair to say they contain the *ends* but not the *ways* and *means*.

Third, working with friends and allies will be imperative in this technologically enabled world. Sharing the results of R&D, as mentioned above, is one important reason to do so. Another is collaborating to define and minimize the risks posed by some dual-use technologies and develop collective strategies to mitigate these risks.

USING TECHNOLOGY TO COUNTER TECHNOLOGY

The democratization and dual-use nature of technology combine to make more capabilities available to a wider array of people who might seek to misuse them. However, this is not to say that someone with malicious intent now has the upper hand.

Constant changes and growing capabilities also imply that counter-measures can be developed to prevent accidents or the misuse of technology and establish response capabilities in case such an event occurs. In some cases, a symmetric technology—biotechnology to counter this use of biotechnology—could be employed. In other cases, asymmetric technologies could be used—for example, use of law enforcement tools to counter development of improvised explosives.

Considering the symmetric case further, biological subelements could be used to counter technologies that had proliferated and could be used in perpetrating an attack. For example, while biotechnology proliferation is making a growing array of capabilities available that could help a bioterrorist with acquiring, processing, weaponizing, and employing biological pathogens, it has also contributed to the development of technologies that would be useful in preventing, identifying, and responding to an attack. Point-of-care diagnostics could assist in more rapidly identifying victims of a biological attack. Bioinformatics could assist in development of medical countermeasures, as well as attribution and forensics should an attack occur. Such medical counter-measures could be developed for treating victims, as could treatment protocols resulting from increased understanding of the progression of disease. But the point is that these technologies must be developed.

In the asymmetric case, consider a social media attack executed by botnets remotely attacking targeted sites or even attacking by dis-tributing false information. The attack could potentially be identified using AI technologies to sense messages coming from nonhuman points of origin. Once identified, these sites could be blocked using the same AI technologies, augmented with traditional law enforcement capabilities and enhanced legal mechanisms to prosecute offenders.

Just as new technologies have been developed in response to human needs, countering dangerous uses of technologies can result in advances as well. Consider the development of cybersecurity, directly resulting from the need to secure the internet against malicious actors and potential attacks.

This section—using technology to counter technology—was not intended to advocate for a technology arms race. Rather, it should serve as a reminder that technology development is an ongoing process, requiring strategic thinking about potential risks (including threats, vulnerabilities, and consequences), dedicated R&D resources to address those risks, and recurring assessments to ensure that risks have been identified and appropriately considered.

CONCLUSIONS

This chapter has examined the modern age of technology, which we have defined as beginning in the post–World War II era. This period has been exciting and dizzying, with extraordinary advances in technological capability beginning with the development of the atomic bomb and ushering in the nuclear era.

In some respects, it seems appropriate to talk about technology before and after World War II, indicating the discontinuity between the two periods. While the Industrial Age beginning in the late 1700s resulted in great change and important advances, the lessons learned providing science and technology for the war effort in World War II fundamentally changed the United States and, arguably, global S&T, R&D, and I&T activities. Furthermore, it resulted in development of an R&D concept that has been foundational and continues to evolve today, both here in the United States and internationally.

The period corresponding to the end of World War II exposed one of the most important questions that continues to confront US R&D today—that is, should the United States have a disconnected model for research? Or should all research be conducted with purposeful outcomes in mind? Or, alternatively, should the United States continue to pursue a blended model that allocates funding to both connectedness and disconnectedness in R&D? And if so, what should the balance be between connectedness and disconnectedness?

The Information Age beginning in the mid-1900s has exponentially accelerated the rate of technological change. It has been accompanied by exponential growth in our understanding of the natural phenomena that guide our planet; convergence in technologies, in which the whole is far greater than the sum of the parts; a democratization of technology, allowing for the proliferation of scientific discovery and technical knowledge throughout the globe; and the diffusion of technology, requiring new methods for obtaining increases in technological capability.

In Part II, we will consider the management of technology based on these changes.

MANAGING TECHNOLOGY

The science of today is the technology of tomorrow.
—Edward Teller

I'm as proud of many of the things we haven't done as the things we have done. Innovation is saying no to a thousand things.
—Steve Jobs

CHAPTER 5

LEADING AT THE SPEED OF TECHNOLOGY

At the intersection of humankind and technology are visionary leaders with the courage to think about the possibilities, take risks in the pursuit of scientific discovery and technological advancement, and fail on occasion. Such leadership has propelled the human race to discovery and advancement fueled by curiosity and built on an ever-increasing body of knowledge.

This chapter considers the importance of science and technology leadership through the lens of three case studies. Through them, one gains a sense of the leadership qualities and attributes that have led to successful technology development outcomes. Just as with leadership in other endeavors, motivating and guiding the next generation features prominently in the responsibilities of today's science and technology leaders.

Equally important for leaders is allocating resources appropriately. This means developing a sense of which avenues to go down and which to avoid. At times, this means becoming opportunistic in developing capabilities for practical purposes by collaborating with others in strategic partnerships to maximize the benefits of one's own and others' R&D spending.

In short, technology development does not simply happen without purposeful effort and the proper allocation of resources. And leadership is a key component of both.

VISION AND LEADERSHIP

Perhaps it seems out of character that a book on technology should include a section on the importance of vision and leadership. However, history demonstrates that both have been critical in many technology advances.

Developing a vision for the future is not too dissimilar from developing a hypothesis, as called for in the application of the scientific method. Based on an idea, concept, or need to fill a practical void, a leader, scientist, or technology developer conceives of a concept for addressing the issue under consideration. This vision or hypothesis then serves as a starting point for investigation. Just as in other endeavors, strong leadership is required to see a technology development project through and ensure that the resources expended on these efforts come to fruition through a successful transition to use.

Three examples will be used to illustrate how vision and leadership have contributed immeasurably to technology development initiatives: the Manhattan Project, putting a man on the moon, and the development of the conceptual and technological underpinnings of the Joint Warfighting force.

The Manhattan Project provides an interesting case study as to how leadership and vision at various levels were instrumental in the development of nuclear weapons. In August 1939, Albert Einstein wrote a letter to President Franklin D. Roosevelt addressing the potential for harnessing the power of a potentially new phenomenon, a nuclear chain reaction. Einstein went on to describe how the discovery could lead to the construction of bombs of a new type with incredible destructive power, such that "a single bomb of this type, carried by boat and exploded in a port, might very well destroy the whole port together with some of the surrounding territory." In this letter, Einstein proposed establishing a program for the development of nuclear weapons, provided thoughts for next steps, and warned that other nations were also likely pursuing such a weapon.[1]

Over a period of three years beginning in August 1942, the United States established the Manhattan Project, which came to employ more than 130,000 people at twenty major sites and cost more than $2 billion. The project, led by Maj. Gen. Leslie Groves of the US Army Corps of Engineers, brought together a distributed, multidisciplinary team with a broad range of scientific, technical, organizational, and administrative specialists from across the United States and in collaboration with Canada and the United Kingdom. It required R&D of new scientific and technology areas including isotope separation, centrifuge technology, electromagnetic separation, nuclear reactors, metallurgy, and weapon design. It also required placing the mission above individual personalities and egos and maintaining the tightest security at all costs.

In December 1942, Nobel Prize–winning scientist Enrico Fermi created a controlled nuclear reaction at the reactor known as Chicago Pile 1 (or CP-1) that served as the proof of concept for the development of a nuclear weapon.[2] Over the course of the next two and a half years, the progress on the largest R&D project in the history of the United States, perhaps even humankind, continued. The material was developed, purified, and prepared to be a component of the nuclear weapon. The trigger mechanisms were refined. The weapons were designed at Los Alamos Laboratory under the direction of the lab's director, nuclear physicist Robert Oppenheimer.

In July 1945, the United States conducted the first ever nuclear test at the Trinity site in Los Alamos, New Mexico. The test produced a yield of twenty thousand tons of TNT. One month later in August 1945, the first atomic bomb was dropped over Hiroshima, Japan, resulting in the deaths of more than 140,000 people. A second atomic bomb was dropped on Nagasaki several days later, causing the death of some 74,000 people.[3]

The scientific, technical, administrative, and organizational leadership demonstrated throughout the development of the US atomic weapons program was instrumental to the program's success. From early scientific and technical leadership provided by the likes of Einstein, Fermi, and Oppenheimer to the organizational and administrative leadership of President Roosevelt and General Groves, all played a decisive role.

By way of postscript, the early years of the atomic program set the stage for establishment of national R&D capabilities in the United States. Without a doubt, the Department of Energy laboratories are direct descendants of the Manhattan Project.

As we will discuss later, the lessons learned concerning R&D support to US efforts during World War II and the Manhattan Project played a significant role in framing postwar science and technology thinking. A key figure in this framing was Vannevar Bush through his report *Science, the Endless Frontier*.

President John F. Kennedy announced on May 25, 1961, before a joint session of Congress that the United States had the goal of sending an American to the moon and returning him safely back to Earth by the end of the decade.[4] The catalyzing speech, corresponding resource commitments, and focused R&D could not have occurred without Kennedy's leadership at this critical moment in history. To be sure, many of the technology building blocks were already in place, but many hurdles needed to be navigated before an American would land on the moon.

Later, in describing specifically why going to the moon was so important and more generally humankind's quest for advancement, Kennedy, in a speech at Rice University, stated,

> We choose to go to the moon in this decade and do the other things, not because they are easy, but because they are hard, because that goal will serve to organize and measure the best of our energies and skills, because that challenge is one that we are willing to accept, one we are unwilling to postpone, and one which we intend to win, and the others, too.[5]

In this speech, he also spoke of Newton and gravity, electric lights and airplanes, and the discovery of penicillin. He reminded the audience of the difficulties of human growth and technological advancement. Through his leadership, Kennedy reenergized a nation and began a new era in the history of humankind.

In the mid-1990s, the chairman of the Joint Chiefs of Staff, General John Shalikashvili, unveiled a document titled *Joint Vision 2010*. In the introduction, the chairman stated, "Joint Vision 2010 is the conceptual template for how America's Armed Forces will channel the vitality and innovation of our people and leverage technological opportunities to achieve new levels of effectiveness in joint warfighting."[6] The effort was centered around five operational concepts: dominant maneuver, precision engagement, focused logistics, full-dimensional protection, and information superiority. The goal of *Joint Vision 2010* was to stimulate the US Armed Forces to take advantage of the opportunities presented by Information Age technologies.

Joint Vision 2010 also served as an organizing construct for resourcing the US military that continues even today. For planning purposes, those technologies and capabilities that were components of or contributed to the five operational concepts were given priority in funding. Using Brian Arthur's concepts to illustrate this integration, the lessons learned in Desert Storm several years earlier provided intellectual grist for thinking about Information Age warfare and what the possibilities might hold. In this way, Information Age tools could be applied to the largely Industrial Age force.

Based on these concepts, General Shalikashvili led an effort to harness the power of the microchip for the advance of warfighting. The new initiative also served to introduce a new organization and resourcing concept called the Joint Requirements Oversight Council, which provided the leadership and technology road maps for developing the joint force into the future. Over the past twenty-five years, the leadership demonstrated by General Shalikashvili has served as the framework for organizing the military and capability development for the US Armed Forces.

BECOMING TECHNOLOGY OPPORTUNISTIC

The need for becoming technologically opportunistic serves an important outcome, as it maximizes the limited R&D resources that are available. Technology opportunism implies strategic scouting of technologies to identify where work is ongoing or may be reaching maturity to allow for incorporation as a subcomponent of a larger technology or even as a stand-alone system.

The goal of becoming technology opportunistic is to link up technology with operators. In a sense, it relies on building relationships to introduce those with the operational requirements to those with access to and understanding of the technology. It also relies on maximizing the allocation of resources to focus spending on those areas that are underserved by R&D but for which there is an important operational requirement.

Market, business, and technology intelligence services are all formal mechanisms for being technology opportunistic, assisting in identifying which companies are working in a particular technology space, how much is being spent on R&D, and which companies or individuals hold the intellectual property rights. Such services can be costly, but they can also be quite beneficial for establishing the state-of-the-art in technologies of interest.

Industry partnerships are essential for being technology opportunistic. Learning where industry has ongoing developmental efforts and what the maturity of those efforts is provides important insights that can assist in technology development and ultimately in solving operational problems. The DoD has become keenly interested in industry R&D, both in terms of gaining access to the results of previously funded internal R&D or IRAD as well as exerting "directive authority over how these companies employed their independent research and development funds."[7] The goal is to get industry to spend its R&D resources to develop next-generation products that align with the highest priorities of the DoD.

Becoming technology opportunistic also means that technologists must be aware of who might be developing such technologies of interest. Doing so could provide an avenue for collaboration that could save on costly early-stage R&D or reduce timelines for technologies that are already being developed. Given the vast proliferation of technology and the number of potential integration opportunities, having technologically savvy strategic scouts who speak the language of technology and understand the use case can be important in the development of a new technology. Such knowledge can also be beneficial for those seeking to upgrade systems; extend technology life cycles; and avoid premature obsolescence of costly, complex technologies.

The DoD makes extensive use of this practice, which partially explains why many of its weapons systems have decades of use. One such example is the B-52 strategic bomber, first fielded in the 1950s, which remains in service today, albeit with nearly complete updates and modernization of most of its subcomponents, from its engines to avionics. Even its payload has been changed from gravity bombs to precision-guided munitions.

Part of being technology opportunistic is strategic teaming with partners to bring in key knowledge, technologies, assemblages, and capabilities. Systems integrators frequently employ this approach. This is not to say that a systems integrator would not have deep understanding and knowledge of how the individual pieces come together. But expecting an individual to have the scientific and technical expertise needed for the development of each of the technology subsystems and their individual technology subcomponents seems unattainable. This is why when leading a multifunctional team, the chief integrator normally has subsection leads who have greater depth and knowledge in their particular technology systems.

For example, many large technology integration companies do not do their own basic research, preferring to focus their efforts on applied research and development and rely on small innovators for more of the earlier-stage research activities. During my time with industry, I was often called upon to review companies' technology products to assess whether they had application for use as subsystems in our technology offerings. These companies were normally medium or small businesses with interesting technologies in the initial stages of development. In some cases, we were looking to procure a final product; in other cases, we were looking for a component technology. During one such evaluation, the company had developed a specialized handheld sensor capability for detecting releases of chemical agents and biological pathogens. The company was hand making their products, and the size of the company did not allow for scaling up to meet envisioned demands. Furthermore, the company understood the science and technology but had no understanding of how to bring the product to market. As a result, their best alternative was either to sell their technology and associated intellectual property rights to a larger company that could commercialize the product or to be acquired by a larger company.

In another case, a company had technology for highly accurate facial recognition that clearly would have been interesting to the Department of Homeland Security. The technology allowed for screening cars at a distance while they moved at a high speed. Unfortunately, the company lacked understanding of the government customers and federal acquisition regulations that would have allowed them to navigate the bureaucracy and offer their technology to the government. In such a case, a technology "Sherpa" or guide could have been highly useful in bringing the technology company to the attention of users looking to solve a real-world problem.

Some of the policies and past behaviors of the Pentagon and other government departments and agencies have made them less attractive to corporations not willing to stand for significant bureaucracy in the form of federal acquisition regulations, low profit margins, and directions on how to spend R&D dollars.[8] This is further exacerbated by the growing demands for technology in nongovernment markets and the increasing democratization of technology. Mending and nurturing these relationships would be essential for government to remain a trusted partner with industry.

In some instances, the government would also benefit by becoming a better consumer. Learning to make use of commercial off-the-shelf technology—where it is appropriate and a high percentage of the requirements can be satisfied with little or no modification—could be beneficial for government. While tailored solutions are likely necessary for the software to run weapons systems, they are likely unnecessary for enterprise software for government financial management and human capital systems. In a related concern, many government acquisition programs (which entail developmental activities) have changing requirements, resulting in increased cost and longer timelines for development and fielding, and jeopardizing the programs' long-term viability.

The Defense Innovation Unit (DIU) (formerly the Defense Innovation Unit Experimental or DIUx) and the establishment of the US Army Futures Command reflect two initiatives designed to make the military a more informed customer and provide greater access to industries' emerging technologies. DIU specifically endeavored to help the US military make faster use of commercial technologies.[9] The Army Futures Command, established in 2018, seeks to support future US Army modernization, specifically by gaining access to new technologies and developing innovative concepts aligned with the army priorities through the use of crossfunctional teams consisting of technologists, operators, and acquisition professionals.[10]

International partners can also be a valuable source of collaboration. In several cases during my service in DHS, international collaboration proved extremely valuable. We frequently shared the results of homeland explosive-detection programs across partner nations. The same was true for information sharing on passenger and cargo screening techniques.

We also benefited from these collaborations. For example, the Netherlands provided important insights to DHS S&T and the Federal Emergency Management Agency on flood control mechanisms. New Zealand shared its work on disaster management and resilience following the massive earthquake in Christchurch several years ago. Australia provided important insights into its Smartgate system used in airports to conduct passenger screening. There were numerous other examples of R&D sharing to the benefit of both parties. In some areas of R&D, history and past experiences implied a partner nation was likely more capable, and the United States stood to benefit greatly from such interactions.

Brian Arthur's core principles—which include the modularity, recursive nature, and hierarchical nature of technology—suggest that over the life cycle of a technology, there may be opportunities to modernize parts of the system to gain either increased effectiveness, efficiency, or longer life. This lead-in also serves to highlight that as an individual technology—perhaps the development of the airplane, as described previously—matures, it might be useful to refresh the technology by upgrading individual components, replace systems that have become obsolete, and perhaps even add new technologies that will increase the functionality of the aircraft. Such upgrades can be found in the F-16 multirole air force fighter.

The original F-16s were fielded in the 1980s, but there are plans to keep them in the force through 2048. Over this almost-seventy-year period, changes would be incorporated to increase the capability of the major systems such that the upgraded model would barely resemble the original. Lockheed Martin, the aircraft's developer, describes the upgrades to the new Block 70 as having "advanced Active Electronically Scanned Array (AESA) radar with a new avionics architecture, and structural upgrades to extend the structural life of the aircraft by more than 50 percent beyond that of previous production F-16 aircraft."[11] The upgraded radars in the aircraft would put it on par with the latest technologies that are now being fielded on the F-35 fifth-generation fighter. The modular nature of the aircraft's subsystems allows for these types of upgrades.

These same concepts work for other technologies as well. For example, cellular telephone networks and individual backbone-system technologies are upgraded when new capabilities become available. The technologies for the network and the cell phones are designed to be backward and forward compatible to allow users to take advantage of the newer capabilities embedded in their cellular phones and the phones to be interoperable with older network equipment. Applications (or apps) that can be downloaded provide users with the ability to manage their own technology and enhance their device's capabilities and the user experience on a personal basis. In this way, users control the technology they wish to have incorporated into their phones. When such technology refreshes occur, the developers normally attempt to make the systems backward compatible. However, after prolonged life cycles, developers are often faced with incurring significant additional expense to extend the life cycle of a technology or to make the earlier technology obsolete.

As technologies continue to be fielded with modern technologies as subsystems, their complexity grows exponentially, increasing the challenges of maintaining and modernizing them. Years ago, mechanics could work on all aspects of the car. Mastering the major components of the car required mechanical skill to repair and replace defective items. Today, the systems are nearly all electronic, requiring specialized testing equipment and knowledge. The number of systems has also increased through the addition of environmental protection, navigation, and electronic ignition.

Becoming technology opportunistic also relies on building multidisciplinary teams that bring together scientific and technical knowledge with operational requirements. For example, bringing together structural engineers, materials scientists, and IT specialists would be important for developing 3-D printing capabilities. Adding operators to the equation provides the opportunity for these technical specialists to develop understanding of use cases they can target with their developmental activities.

At times, R&D feels quite stovepiped, with little communication between different communities or between users and developers. Data scientists are often working to develop new ways to identify the signal within the noise, yet they lack actual data sets on which to test their data. Chemical engineers may be developing technologies for detecting homemade explosives and may not understand the constraints of the operational environment in which their systems or algorithms will be required to operate. Designing a seven-by-twenty-square-foot stand-alone imaging machine may not be useful (or feasible) if the airports only have room for a three-by-four desktop imager.

Such cross-fertilization can have big benefits in reducing the need for R&D for products that can easily be integrated into other systems. Consider the navigation and technology packages in today's automobiles. Certainly, they are orders of magnitude more capable than even several years ago; however, the car companies have attempted to design their own interface packages that are far less capable than what is on today's smartphones. Would they be more efficient and improve customer experience if they were to integrate a smartphone capability directly into their new car platforms? In fact, we are beginning to see more vehicles being developed that will provide direct interface with either Apple or Android cellular systems.

Cross-fertilization can also apply to bringing disparate groups of technologists together at critical times to review R&D that is ongoing. It can be interesting to have physicists reviewing cybersecurity algorithms or data scientists looking at biotechnologies. Unexpected opportunities surface when these kinds of cross-fertilization occur. This should not be a surprise, as many of the breakthroughs in various fields come from the combination of other technologies.

As an example, bioinformatics is based on the integration of information technology, data science, and biotechnology. The Human Genome Project, which started in the early 1990s, began to employ early applications of what we now refer to as bioinformatics. Increases in information technology allowed for faster computing and more accurate sequencing of the genome. New data algorithms contributed to more readily determining relationships in areas such as relating disease to specific genes or genomic sequences. In looking to the future, as greater biotechnology knowledge is acquired, new uses will surely be developed. For example, bioinformatics will undoubtedly contribute to the development of models and simulations that will assist in a wide variety of applications, from rational drug design to understanding the progression of disease at the cellular level. Without these combinations of technologies, it would not be remotely possible to even conceive of gaining an understanding of how the 3.3 billion base pairs of human DNA function.

Finally, being technology opportunistic implies thinking more broadly about developing operational capabilities. One method for doing this is the structured DOTMLPF—doctrine, organization, training, material, leadership, personnel, facilities—framework that focuses on building capabilities more broadly using S&T, R&D and I&T methods. The framework was originally used in DoD but is now employed more broadly across the US government. The implication is that technology development could result from building new tools and applications or through process reform and changes to organizations.

THINKING DIFFERENTLY ABOUT TECHNOLOGY DEVELOPMENT

As previously highlighted, conducting R&D, especially in the initial stages of research, can be extremely costly and result in only marginal increases in performance. This relationship is the essence of the S-curve model of technology development. One technique for conserving resources and becoming more technologically opportunistic is to go from R&D to "little r, big D." In this paradigm, one would rely on others for the bulk of the research and focus on acquiring technologies that could be developed for the desired operational use case.[12]

A basis for this r&D approach was market analysis (sometimes called technology foraging) using commercially available, industry-specific technology databases (such as the market, business, and intelligence databases discussed previously) that can provide insights into a technology's development and maturity. This approach also relies heavily on building partnerships across the government, with international collaborators, and with industry. By sharing insights and pooling resources, a synergistic effect can be achieved. For example, DHS S&T and the Environmental Protection Agency had a partnership involving determination of standards for decontamination in the event of a biological attack. In another example, DHS and DoD had a broad range of collaborative efforts, from tunnel detection capabilities to biometrics, that proved to be mutually beneficial.

A r&D approach does imply limiting internal resources for basic and applied research while focusing funding on technology development. However, such a strategy does not imply that no research will be done but, rather, that it will be done sparingly in key areas where others are not working or on specific operational issues. This technique was used with some success during my time in DHS S&T. Areas such as development of algorithms for detecting homemade explosives garnered little interest from others, outside of a narrowly focused law enforcement and defense community or commercial industry looking to employ these algorithms in systems that would find their way into airports and federal buildings. These areas received research funding through DHS S&T due to their unique nature.

The r&D strategy also implies that other strategies will be pursued for gaining the benefits of the research being done in the broader R&D community. Resources could be maximized by partnering with others. In this way, the research risks and benefits could be shared between partners. In such cases, legal instruments should be employed to assure that all parties understand their respective roles and potential rewards. One area where partnering was employed was biodiagnostics, which potentially could have a large market, even one that would dwarf the DHS's needs for use in biosurveillance. In partnering with industry and academia, the R&D became a shared program and responsibility.

CONCLUSIONS

Leadership—just as in other human endeavors—is an essential part of technology development. Having a vision for the direction of the technology development and applying leadership to R&D efforts are important for achieving outcomes with the appropriate balance of near-term problem-solving and longer-term understanding.

Becoming technologically opportunistic by capitalizing on the technology development efforts that have been undertaken previously and incorporating the results of their progress into your work can be a resource-saving strategy.

Finally, technology development does not normally fit into a linear model where one proceeds from step to step in an ordered fashion. While this might be the preference for efficiency, in reality, one should expect considerably more nonlinearity as new discoveries are made, lessons are learned through experimentation, and new subtechnologies are developed and incorporated. This nonlinearity implies that the technologist must have a degree of agility and flexibility that promotes a technology development process prepared to take advantage of the opportunities that are encountered through discovery and mitigates the challenges that must be overcome.

CHAPTER 6

ASSESSING TECHNOLOGY

Invariably, technologists are asked to look into the future and describe the environment they envision. What technologies will dominate in the next five, ten, twenty-five years? How will society and the lives of individuals change as new technologies are incorporated into societies and into the lives of individuals?

Some technology market research firms, specializing in looking into the future, have developed capabilities for considering likely changes in fields of technology. Many use combinations of proprietary data, methods, and algorithms to assess future market trends. Some of the market research firms are highly specialized and therefore focus only on a single technology field, while others advertise being able to look broadly at numerous technology fields. Many even allow comparisons across different fields of technology.

No doubt a great deal of data about the technology field's history and successes will be contained within these qualitative data. Unfortunately, these data are looking at past information and attempting to make predictions about the future. Such techniques are interesting, but may not necessarily assist in looking to the future at an individual technology or field of technology.

In this chapter, a framework for looking into the future will be offered to the reader for assessing individual technologies or technology fields. The concept will rely on looking to the future to assess how the interplay among five attributes—(1) science and technology maturity; (2) use case, demand, and market forces; (3) resources required; (4) policy, legal, ethical, and regulatory impediments; and (5) technology accessibility—will combine to affect the technology's development.

THE INSPIRATION FOR OUR TECHNOLOGY ASSESSMENT FRAMEWORK

One prescient evaluation of technology provides a historical perspective on biotechnology, looking specifically at those technologies which could provide dual-use capabilities for the development of biological weapons. The analysis was conducted in 1998 by the Congressional Office of Technology Assessment and examined the rate of change of fourteen biotechnologies over the sixty-year period from 1940–2000.[1]

While the analysis did not make any assertions about the future, it contained a chart showing that prior to 1970, most of the technologies were developing at a rate where the capacity was doubling every five years. In the 1970–1980 period, this rate increased such that the doubling of capacity was occurring annually. Beginning in 1980 and continuing to the end of the assessment period in 2000, the rate of change for all the technologies was doubling every six months.

By the end of the period considered in the analysis, the technologies were increasing at a rate of 400 percent annually. Many, such as monoclonal antibodies, antibiotics, cell growth chambers and fermenters, the Human Genome Project, and DNA engineering, have since reached a level of even greater maturity over the twenty-year period since the study was conducted.

This simple graphic raised several important questions. What's on the other side of 2000? Will these technologies continue to increase at this pace, or will they soon reach the top of their S-curves and lead to other technologies that will inevitably follow the same exponential development paths? And what are those new technologies that are just beginning their climb up the S-curve? If we were to redo the previous analysis, how can we even begin to pick the technologies that would be on the other side of 2020?

The very act of selecting the important technologies requires an ability, perhaps a special talent, for looking into the future to see what lies two terrain features out. In other words, this analysis should not just consider tomorrow but tomorrow's tomorrow when looking at future technologies. It requires developing the capacity to think multi-generationally in examining those technologies likely to dominate, how they are likely to evolve, and how they are likely to shape their respective fields and even humankind.

When Gordon Moore of Intel Corporation observed in 1965 that the number of transistors on a square inch of integrated circuits was doubling every two years, the importance transcended the changes related to the physical size of the integrated circuit. In effect, the physical size became a shorthand to describe the increasing capacity to build more powerful microchips to handle increasingly complex systems, mathematical calculations and computer languages and code, as well as to allow smaller footprints, which ushered in the era of mobile computing. And it also saw the reduction of cost per operation as the industry matured, making these capabilities increasingly available to all. Today, the two-year mark for doubling the number of transistors has been reduced to eighteen months, yet some are now predicting that in the 2020s, Moore's law might reach its physical limits.[2]

All of this is to pose the question of what is on the other side of Moore's law. What does tomorrow's tomorrow look like as we look two terrain features into the future?

LOOKING TWO TERRAIN FEATURES INTO THE FUTURE

The structured use of forecasting methods dates to the end of World War II. A 1945 report prepared for the US Army called *Toward New Horizons* sought to identify key technology changes that occurred in the war and make recommendations about future R&D areas that might be pursued. In the 1960s, the RAND Corporation developed the Delphi method, which was a structured way to consider difficult questions that involved asking experts and narrowing down opinions until consensus judgments could be identified.[3]

These efforts involve using the knowledge of the past to predict the future. In short, they look to the past to understand what occurred and use this knowledge to predict future trends. In the cases cited above, the analyses have been constrained from the beginning. In the first case, historical precedents from World War II serve as the starting point, while in the second case, a method that relies on expert opinions from those with deep functional expertise about individual technologies—yet likely little expertise in thinking as a futurist—would make judgments about the convergence of technologies in the future. To get the best of both worlds would require subject matter experts with deep knowledge of a technology field coupled with a broader understanding of technology development, including the convergence of technologies and what this portends for the future.

The potential for generating value for these predictions would be to see into the future and make bets on certain technologies and products. This could be useful for governments, industry, and even individuals. Governments could employ this knowledge in crafting strategies and policies that seek to take advantage of opportunities and mitigate challenges. Industries could use such knowledge to drive technology development. Those who fail to read the road signs are apt to pay by losing market share or becoming obsolete. Armed with this information, even individuals could be better prepared to make informed decisions about incorporating new technologies into their daily lives.

The same concept holds for technology development for militaries, where investments in technology and the integration of systems of systems can be pivotal for designing deterrence, defensive, and warfighting capabilities. Yet failure to read the signs and adapt to the changing technological landscape can be catastrophic, as the example of the Maginot Line demonstrates. So how are these predictions done today, and could a method be designed to more clearly focus on the future rather than relying on the past?

In evaluating technology, one often does side-by-side comparisons. The military uses this approach to compare its fighter aircraft to those of a potential adversary. For example, such an analysis might entail comparing the United States' F-35 Lightning with China's J-31. Both are stealth multirole aircraft, considered to be the most modern for each nation. However, comparing attributes such as range, weapons, avionics, and speed can be misleading when trying to understand the overall contribution to warfare.

In fact, comparison of these aircraft should be less about the individual systems than how they are combined with other key systems to produce battlefield effects. The capabilities of the individual aircraft become far less important than the air battle management system in which they are operating. According to US Air Force doctrine, the individual aircraft are flown by pilots who receive intelligence and direction from an airborne warning and control system (AWACS) that provides "airborne surveillance, and command, control and communications (C3) functions for both tactical and air defense forces."[4] AWACS provides the output of powerful standoff radars that can be relayed to the individual aircraft, thus affording greater situational awareness for the pilots in the fighter aircraft. The incorporation of aerial refueling tankers increases the operational range of the aircraft, thus allowing the fighters to stay airborne longer, cover more distance, and remain in the fight for more time.

Just as in the earlier example of the development of aviation technology and the related systems of systems including the air traffic control, one must take a broader view in comparing relative strengths and weaknesses. In this way, it is the combination of technologies— really the integration, coordination, and synchronization of the technologies—that contributes to an overall warfighting capability. This means that individual comparisons of aircraft can be interesting, but they only tell part of the story and—if not considered within the operational context in which the aircraft will be employed—can even be misleading. The real comparison should be assessing the comparative offensive and defensive employment of systems for the United States and potential adversaries to gain an understanding of which side has the advantages.

What we are really looking to identify are forecasting techniques that can provide insight into the future. As you might imagine, a variety of such methods have been developed. For example, one source identifies four methods: judgmental or intuitive methods, extrapolation and trend analysis, models, and scenarios and simulations.[5] Of course, these methods could be used individually or in combination when considering futuristic outcomes.

The judgmental or intuitive methods depend on qualitative techniques involving the elicitation of experts. The Delphi method discussed previously falls into this category. Often, a first step in forecasting is to convene experts to at least set the historical landscape to describe how the present has evolved. Once the current state is understood, the experts can be queried, individually or in a group, to gain insight into how they see the evolution in the future.

Extrapolation and trend analysis are similar to the biotechnology assessment presented earlier. If one were to predict what was on the other side of the year 2000, list the technologies, and assess their likely rate of increase, one would be employing extrapolation and trend analysis. Considering the technology's S-curve and relating it through analogy to other technologies' developmental profiles can provide insights into how the technology in question might evolve in the future.

One such method is called TRIZ or Theoria Resheneyva Isobretatelskehuh Zadach, which is Russian for "inventive problem-solving." TRIZ is based on the principle that your problem or one similar to it has already been solved at some point. Thus, creativity means finding the earlier case, adapting that solution to your issue, and resolving all the contradictions. The deployment of airbags in automobiles provides two such contradictions to be considered. The systems must deploy quickly, but if they deploy *too* quickly, they can kill or injure someone. They must also be stored in a compact configuration and, when deployed, expand to many times their original volume.[6] An inherent assumption in TRIZ is that future technology will evolve in much the way that technologies have evolved in the past; this sounds similar to the evolutionary technology theory that Brian Arthur developed and in practice contains similar pathways for thinking about the nature and future of technology.[7]

Models can offer a variety of techniques for developing conceptual and mathematical representations of technology futures. One such model uses an economic analysis centered on understanding rates of return to assess future technology trends; this model is quite similar to the use of the S-curve in determining where the technology falls within its life cycle. Another model uses chaos theory and artificial neural networks to look for repeatable patterns in previous technology development that can suggest the future for the technology being evaluated. The use of influence diagrams provides yet another model that has been identified for technology forecasting; such techniques allow one to determine the logical cause-and-effect relationships between related activities and the probabilities of those outcomes. Tools such as the critical path method and PERT (Program Evaluation Review Technique) provide specialized tools for considering such relationships and have proven useful in project management, of which technology development can be considered a subset.

Scenarios and simulations can also provide insights into future technology development. However, just as models are representations of reality, so too are scenarios and simulations. These tools normally rely on experts to participate in the forecasting. Another interesting tool in this category is backcasting, which entails presenting a future outcome and examining the path taken to get there.

As this section has highlighted, various techniques have been suggested for forecasting technologies. Several thoughts come to mind in our quest for developing a technology forecasting methodology. First, the various methods—judgmental or intuitive methods, extrapolation and trend analysis, models, and scenarios and simulations—have long histories in analytical endeavors going back decades. Second, none of the methods in isolation will likely be sufficient to conduct technology forecasting. Third, accurate technology forecasting will not be based on examining individual technologies or even technology families but, rather, on how various principles, methods, and concepts are combined to develop a broader conception of the development—and convergence—of future technologies.

CATCHING THE NEXT BIG WAVE: DEVELOPING A FRAMEWORK FOR TECHNOLOGY AVAILABILITY FORECASTING

The point of departure for our technology forecasting framework will be the assertion that five areas can be considered to gain an understanding of the prospects for a technology in the future: (1) science and technology maturity; (2) use case, demand, and market forces; (3) resources required; (4) policy, legal, ethical, and regulatory impediments; and (5) technology accessibility.

The five areas have complex interrelationships that can, when considered in combination, make the technology either more or less available to a potential user. If the science and technology maturity is low but there is a strong use case and market demand, key resources could be employed that could accelerate the S&T efforts and potentially reduce technology development timelines.

For technologies that are mature with high demand and adequate resourcing and with few or no impediments to use, either formal or informal, the technology would be readily available for users. Alternatively, for technologies that are not mature or that are limited in some way, the technology would be less readily available. The concept of availability also provides insights into the level of skill that could be required to use a technology.

Our Technology Assessment Methodology provides a structured approach to considering technology availability. For each framework area, assessment criteria guide the analysis of a technology. Each area is composed of five components, an overarching assessment question, and levels that define the assessment scoring. Using a decision support scoring methodology, the technologies can be assessed. Appendix C provides a graphical depiction of this framework and a more detailed description of the scoring methodology.[8]

SCIENCE AND TECHNOLOGY MATURITY

Science and technology maturity provides a basis for considering the history, maturity, and rate of change of the technology. The establishment of the TRLs provides an understanding of the life cycle progression of the technology. Is the technology in research or development? Have prototypes already been developed and tested in an operationally relevant environment? Have supporting technologies—the components of the system—reached the appropriate level of maturity to be included in the technology being developed?

Related to TRLs are knowledge readiness levels (KRLs) and manufacturing readiness levels (MRLs). KRLs can be very useful in assessing whether the technology area is sufficiently understood. Are the physical phenomena being employed well understood, or is further research or early development required to gain a fuller understanding?

MRLs correspond to the producibility of the technology. Understanding the MRL is critical for a technology to cross the valley of death and become operationally relevant. Can the technology be produced in quantities needed to have an operational impact, or can it only be produced in small quantities? Does the product depend on component technologies with long lead times? Is there a viable supply chain for the technology to be produced (including any raw materials such as rare earth metals)?

Publications can also provide insights into the maturity of the science and technology. When an emerging technology is in early research, publications are likely to be limited to research papers and dissertations. As the technology matures, early adopters will begin to publish information about the potential use cases and any breakthroughs that have been achieved. Some of these early publications will likely present information on potential use cases that might be on the horizon. As the technology matures, the publications will expand to include more mentions in applied research and technology development journals, where secondary or complementary use cases will begin to be highlighted. As the technology reaches maturity, market integration through commercial offerings will dominate the publications.

Patents provide important protections for intellectual property. The requirements for filing a patent and gaining approval for this protection imply a level of maturity at which the technology can be defined and meets established criteria for patentability. These criteria include that the technology can be covered under patent law; is novel or providing a newness or innovation; and is nonobvious, meaning it is distinct from previously awarded patents. Given that the term of grant of a patent is twenty years, reviewers seek to ensure a patent that protects the inventor without establishing overly broad property rights that could infringe upon other innovations related to a technology field.[9]

Miniaturization provides a measure of the maturity of the technology. Perhaps this is best illustrated using an example, the cellular telephone. One need only compare the earlier cellular systems, which were larger and heavier than a brick, with the small, lightweight systems of today to gain an appreciation for the importance of miniaturization (and market demand created through this miniaturization) in determining the maturity of the technology. Miniaturization signifies not only whether users are likely to want to adopt the technology but also the likelihood that the technology will be included as an embedded technology within another technology.

Life cycle improvements also signal maturing of a technology, perhaps even an aging of the technology. Again, the cellular telephone provides a relevant example—the smartphone has been through five generations in a ten-year period, signifying its continued life cycle development. In some cases, multiple life cycle changes could imply a technology that might be on the verge of being replaced by a newer innovation with greater capability, perhaps even in a smaller package or footprint.

To assess this area, one would consider the maturity by answering the following question: "What is the level of maturity of the science and technology under consideration?" By considering the five components that define this area, a technology under consideration could be assessed according to the specific rating levels that indicate the maturity of the technology and therefore the effort that could be required to bring it to maturity. The levels are (1) elemental, (2) proof of concept, (3) component and system validation, (4) full prototype, and (5) mature technology.

USE CASES, DEMAND, AND MARKET FORCES

Use cases, demand, and market forces define how the technology is likely to be used, whether it is likely to cross the valley of death, and what demand there will likely be for the product. Will the technology likely be a core technology or building block for advancement in other technologies?

The five components of this technology area are perceived investment risk for technology development; need or potential impact of innovation; number and types of people or groups using the technology; business case; and number and size of companies that own IP or are doing the R&D.

Perceived investment risk for the technology development is related to the current level of maturity, seeking to determine the investment required to develop the technology into a commercial product. This also relates to the business case (which will be discussed more later)—that is, should the investors see the commercialization of the technology as a sure thing and likely to provide a good return on investment (ROI), or is the technology so immature that significant investment will be needed and the demand is not known at this point?

Need or potential impact of the innovation identifies the likely use of the technology for first-, second-, and perhaps third-order technology systems. Will the technology be so compelling that it will be necessary for everyday life? Is the technology likely to become a component of other technologies? And if it is likely to become a component, will it see wide usage?

The number and types of people and groups using the technology signifies where the technology currently is in its life cycle. Is the technology so specialized that it is not available without tacit knowledge, great expense, and specialized facilities? Combined with the earlier assessments of the science and technology maturity in the first area (including TRLs, KRLs, and MRLs), one could determine how likely a technology is to become widely adopted throughout the market. Will market forces propel the technology, or will developers and investors struggle to attract customers and need to stimulate demand and pull the market? Is it likely to become an essential part of everyday life or a technology that only a few specialized users will require?

The business case refers to the likely ROI to be generated. What is the initial investment? When is it likely to be recovered? What is the anticipated risk of the technology development? Is the technology a sure thing, or is there some uncertainty? The business case also relates to factors such as market share that are likely to be captured once the technology has reached maturity.

The number and size of the company (or companies) that owns the IP or is doing the R&D matters. Is the technology being developed by a small number of people who essentially are the horsepower behind the technology? Or are there likely to be issues with gaining access to the IP? Is the ownership clean, or are the relationships between the companies holding the IP and those doing the R&D complex and even uncertain? Are any foreign investors involved that could complicate IP protections? Are the companies commercial entities or small R&D firms?

Assessing this area would be done by asking, "Is the technology likely to be an essential component or building block for advancement in biotechnology or other fields?" Using the five components that define this area, a technology under consideration could be assessed according to the specific rating levels that indicate the use case, demand, and market forces. The levels are (1) innovator use only, (2) early adopter use, (3) specialized uses only, (4) majority users, and (5) broad use in a variety of applications.

RESOURCES COMMITTED

Resources committed can provide useful metrics for considering the investment spending and thereby the potential for successfully developing a technology. The five components of this area are indirect investments, technology development costs, total research dollars allocated to the technology, organization of the investment, and related grants.

Indirect costs are critically important to the development of technology. Requirements for specialized facilities, laboratory space, equipment, and personnel can be important cost drivers that must be factored into technology development costs. In some cases, elemental or initial R&D may be required to produce specialized systems to assist in the development of the technology.

Estimating the total technology development costs is also essential and speaks volumes to the likely success of the endeavor. For example, if a technology is in the early stages of development where basic principles must be investigated, yet the resources committed only address a small portion of the effort, this implies that additional resources will be required to address the full scope of the technology.

Understanding the cost structure of the technology development project is also imperative. What are the total development costs of the project and the total research dollars available? A project could look to be feasible until one realizes that the development costs are equal to the research dollars before factoring in overhead requirements. Are additional funds, such as through internal research and development, required?

Understanding the organizations making the investment is also imperative. Is there only resource commitment for a limited or short-duration investment? Have they contracted for a multiyear project, but they only get their funds annually, which means that all funding beyond the first year is at risk? Are the investments tied to a for-profit business?

What are the sources of funding? Is the organization funding the project government, industry, or a foundation? Is the technology development covered by grant funding? Are there constraints on the funding that could limit the type of work conducted and how the results of the work can be used? For example, a grant funded by a nonprofit might not want the funds to be used to develop commercial technologies. Such an arrangement might even jeopardize the organization's nonprofit status.

To assess this area, one would consider the resources committed by answering the following question: "Will resource constraints serve as a barrier to development of the technology?" Using the five components that define this area, technology resource investments could be assessed to be (1) little or no investment, (2) declining investment, (3) steady investment, (4) increasing investment, or (5) exponentially increasing investment.

POLICY, LEGAL, ETHICAL, AND REGULATORY IMPEDIMENTS

Considering policy, legal, ethical, and regulatory impediments allows one to examine the potential barriers that have been or are likely to be imposed. The five components of this area are policy, legal, or ethical barriers; export controls; technical leadership; existing regulations and ability to regulate; and comparative advantage.

Do any barriers currently exist to the development and fielding of the technology? To what degree will this technology be considered dual use and therefore likely to be marked for special handling? Could the technology have applications in the development and use of weapons of mass destruction? Do any arms control agreements (or other international arrangements) relate to the technology? For example, commercial aircraft developers would like to develop large unmanned aerial systems for transporting cargo; however, such a platform would violate the Missile Technology Control Regime (MTCR), which limits ballistic and cruise missiles to payloads of less than five hundred kilograms and a range of less than three hundred kilometers.

Policy impediments could include approval by entities, such as institutional biological committees in the case of biotechnology. In DHS, all biological experiments conducted in our high-containment labs required approval through the department's compliance review group, which certified compliance with the Biological Weapons Convention. This could certainly slow down or even prevent an experiment from being conducted. Are policy barriers in place that could hinder technology development, such as for GMO and stem cells?

For legal barriers, is there likely to be any intersection with arms control treaties or other international laws? Are national laws in place that would limit or constrain R&D? Will special legal provisions be required before R&D can continue? Are there patents in place that could constrain technology development?

Export controls can be severely constraining in some cases. Is the technology covered under export control under either the Department of Defense's US Munitions List or the Department of Commerce's Commerce Control List? Will foreigners be working on the technology development, in which case, if the technology is export controlled, licenses would need to be obtained to allow them to work on the project?

Examining technical leadership can provide an understanding of how important international leadership is likely to be should an export license be required. Does the technology development depend on specialized equipment or facilities, which could affect timelines for maturing the technology and even the feasibility of the technology development should access not be available?

Existing regulations and ability to regulate can also have an impact on technology maturity timelines. Do existing regulations constrain development in some way? Does the technology fit within existing regulatory frameworks? Consider if the Federal Aviation Administration had severely restricted flight operations in the early days of drones. This could have prevented the technology development and the demonstration of the utility of these systems for a wide variety of use cases.

Determining comparative advantage can also help to identify which companies or even nations might have an advantage in the development of a technology. For example, green (or environmentally friendly) policies by a nation might give one of its companies a comparative advantage in the development of energy-efficient batteries or renewable energy systems. In the same way, assessing comparative disadvantages can assist in identifying technical barriers that might be present. Is the comparative advantage favoring smaller, nimble start-ups or larger commercial corporations? Does government or industry have an advantage in the development of a technology?

To consider this area, one would assess the policy, legal, ethical, and regulatory impediments the technology might face by answering the following question: "To what degree are policy, legal, ethical, or regulatory barriers likely to serve as a barrier to the development of the technology?" Using the five components that define this area, the barriers could be assessed to be (1) absolute, (2) significant, (3) some, (4) few, or (5) no barriers.

TECHNOLOGY ACCESSIBILITY

Accessibility allows for considering the technology from the point of view of the user. At first glance, this area might appear to overlap with several other areas. However, these other areas consider the technology development progression and capabilities, while this area considers the technology from the user's perspective. From this vantage point, they are very different indeed. The five components of this area are cost to use, regulatory, technical, access to the technology, and democratization and deskilling.

Cost to use requires assessment of the recurring costs to the user of the commercial technology. Is it prohibitive? Will there be costs other than the financial that might make use of this technology less beneficial? For example, if a very efficient technology were developed, but it created significant environmental concerns, would the user shy away from using it? Are specialized facilities required to operate the technology?

Previously when we considered regulatory, we addressed regulatory issues associated with conducting the R&D. In this case, the regulatory issues pertain to the use of the technology by a customer. Are there existing laws that could prevent or limit use in some way? Does the operation of the technology need to be regulated based on ordinances? Does it create noise and vibrations that are unacceptable?

Technical relates to whether the technology requires great skill or intensive training to use. Is tacit knowledge required for a user to usefully employ the technology? How much training and education would be required? How dangerous is the technology for the operator? Or are basic mechanical skills, modest training, and basic safety measures adequate for use of the technology?

Access to the technology refers to the general accessibility of the technology. Is it only sparsely available, or has it been distributed widely? Can it be procured online, or is it difficult to obtain?

Finally, democratization and deskilling refer to a more general way of assessing the technical feasibility for an end user. Could an end user not only employ the basic technology but also use it as a component in a self-designed system? Information technology provides such an example. Consider that all components of the network are readily employed individually but could easily be reconfigured for another use case.

To assess this area, one would consider the technology accessibility by answering the following question: "Are the controls on the technologies likely to limit access to the wider population or be too technically sophisticated for widespread use?" Using the five components that define this area, the technology accessibility to the user could be assessed to be (1) extreme, (2) significant, (3) some, (4) few, or (5) no user accessibility impediments.

USING THE TECHNOLOGY ASSESSMENT METHODOLOGY

Having defined the five areas (i.e., [1] science and technology maturity; [2] use case, demand, and market forces; [3] policy, legal, ethical, and regulatory impediments; [4] resources required; and [5] technology accessibility), we can now develop a methodology for a structured technology assessment that combines them into a single framework. Using this framework, individual technologies or technology fields can be assessed across the five areas using their respective components. The components add structure to the assessment; however, the methodology also has the flexibility to add or subtract components depending on user preferences and the specific technology under consideration.

The questions that served to frame the assessments provided a basis for developing a normalization process that allows technologists to compare the results on a common scale and make judgments regarding future technology trends. For example, for the science and technology maturity area for a technology under consideration, the technological maturity would be graded according to the five rating levels: (1) elemental, (2) proof of concept, (3) component and system validation, (4) full prototype, and (5) mature technology. Evaluating each of the five components for an area—in this case, the science and technology maturity area—allows one to look more closely at the maturity, relying on many factors in conducting the assessment.

Literature searches, business intelligence, and combinations of techniques presented previously (i.e., judgmental or intuitive methods, extrapolation and trend analysis, models, and scenarios and simulations) could all be used to gain insights into framework areas and their components, the state of the technology, and the potential for transition and commercialization.

This same method would be repeated for each of the other four areas: use case, demand, and market forces; policy, legal, ethical, and regulatory impediments; resources required; and technology accessibility. The scores can be combined using decision support methods, which will provide a basis for comparing each of the five areas. Areas can be assigned specific weights if those assessing the technology believe some are more important to the assessment than others.

Selecting technology categories needs to be done carefully so that accurate assessments can be made. If a category is too broad, the assessment will likely be inaccurate and perhaps even less useful. For example, nanotechnology could potentially be used in maritime coatings, for medical purposes (such as the delivery of drugs to diseased cells in the body), in electrical circuits due to superior conductivity properties, or in pollution control. The varied number of uses for nanotechnology suggests that perhaps it would likely be better to consider the individual use cases rather than the overarching technology.

Each use case will undoubtedly have a different evaluation across each of the five areas. Nanotechnology for use in coatings has far greater technological maturity and would certainly require less regulation than use of nanotechnology in medicine for invasive delivery of drugs within the human body. In fact, nanotechnology applications for coatings have been employed for over a decade, while those invasive medical applications are likely a decade or more away.

This is not to say that there would not be utility to more broadly looking at the evolution of nanotechnology as a category in some cases. However, such a decision would be based on the type of comparison that one is attempting to undertake.

So how might this technology assessment framework be employed?

First, an individual technology can be examined over time to assess how it has evolved and is likely to evolve in the future. For example, in one analysis I led at RAND, specific biotechnologies were examined in 2000, 2018, and 2030 to look at the past, present, and future periods. By examining the past and current trends, one could extrapolate into the future, in this case out to 2030.

Second, examining an individual technology in this way could be useful in determining likely teaming partners for R&D projects. Understanding the relative advantages that each performer might bring to the R&D project could increase the potential for success.

Third, this methodology could facilitate comparing various competing technologies to arrive at an assessment of which technology is most likely to complete a successful R&D process and transition.

Certainly as the reader gains familiarity with the Technology Assessment Methodology, other applications are likely to be identified.

CONCLUSIONS

Technologists invariably are asked to investigate the future and describe the environment they envision. Which technologies will be important in the future? Will a technology be foundational or merely a "flash in the pan" to be quickly replaced by other, more powerful technologies? How will a new technology or field of investigation affect society? And which are the technologies most likely to be disruptive?

Accurately assessing technologies is an important skill for the technologist to master. To aid in this analysis, throughout history, numerous tools and methods have been developed. In this spirit, the Technology Assessment Methodology explained in this chapter adds another such capability, which hopefully the reader will find useful in examining the evolution of technologies. The methodology provides a straightforward and comprehensive approach for considering the past, present, and future development of individual technologies and provides a means to compare different technologies.

Furthermore, the Technology Assessment Methodology seeks to incorporate the broad concepts we have discussed earlier regarding technology development, from readiness levels to forces that will contribute to or detract from further development of a technology.

CHAPTER 7

TECHNOLOGY DEVELOPMENT METHODS

Methods for managing technology development have evolved over the course of human development; however, they have only become part of structured technology development practices over the past seventy years. This structure coincides with the development of formal concepts for research and development. As we have discussed previously, early technology development tended to be less formal and more decentralized. However, armed with the lessons learned in World War II, a more systematized approach to R&D was developed, beginning with the reliance on stated operational requirements.

Establishing operational requirements is an essential component for developing technology programs that are connected to real-world problems and therefore more likely to successfully transition to use. The system in use throughout much of the government today is called the Planning, Programming, Budgeting, and Execution System (PPBES) and serves as a mechanism for alignment and measurement of programs to determine their linkages to strategy, operational requirements, and resource expenditures. While the PPBES normally is discussed in relation to a department or agency's broader strategy that begins with high-level goals and objectives and ends with a budget, the same principles apply to the individual programs, such as those related to R&D. Therefore, one can look at the overall R&D portfolio to assess its health or, alternatively, examine the individual R&D programs within it.

Systems such as the PPBES and TRLs also provide control mechanisms for technology development. Both provide structured ways to describe the progression of technologies across their life cycle. In the case of TRLs, the technology maturity is described, while for PPBES, the life cycle resourcing is the focus.

By design, this chapter stops after the R&D process and therefore does not move into the realm of acquisition. However, to be fair, the line between those two areas is not so clear. Many of the activities conducted as part of R&D—especially those in the later stages beyond TRL 5—can really be classified as part of the acquisition system, or certainly as risk-reduction mechanisms prior to acquisition. Consider prototyping, which occurs in TRL 7. In procurement today, prototyping has become a permanent fixture, especially for large, costly multiyear procurements.

An inherent goal of technology development efforts since the 1950s has been to increase the communication and collaboration between operators and R&D performers. Numerous formal and informal programs have been established to promote this dialogue. Despite these efforts, this area remains one of the most important challenges facing the S&T, R&D, and I&T processes today. Still, it remains a necessary component of purposeful technology development.

McNAMARA AND THE PPBS

The predecessor to the PPBES was the PPBS, which stood for Planning, Programming, Budgeting System.[1] At its core, PPBS was about making strategically informed and structured resource decisions. Such decisions should include the priorities and funding for S&T, R&D, and I&T.

PPBS has been credited with being an integrated management system, a systems analysis method, and a decision-making tool. It was first introduced in the 1960s under Secretary of Defense Robert McNamara. The RAND Corporation, an FFRDC formed in 1947, has been credited with being the intellectual heft behind the development of PPBS.[2]

When Robert McNamara became secretary of defense under President Kennedy, he found himself entering into a very different world than what he had experienced at the Ford Motor Company. He was managing arguably the largest, most complex department in the US government at a turbulent time in our nation's history. World War II had demonstrated, and the National Security Act of 1947 had proclaimed, America's changing role in the world. Despite a large budget, the DoD had no structured way to consider budget alternatives, including R&D expenditures. Even worse, the military planning and financial management of the department were being considered separately. The military services—US Army, Navy, Air Force, and Marine Corps[3]—were responsible for organizing, training, and equipping their forces, while the Joint Chiefs of Staff were responsible for conducting planning for the employment of forces. This separated the resourcing and operational planning functions into two stovepiped processes. Such a disconnect could result in the development of capabilities for which no operational plan existed—in other words, building capabilities with no operational requirement.

Experiences in and since World War II had also confirmed the importance of two key areas, systems analysis and the connected model of conducting R&D. The lack of a Pentagon management system that would link operational requirements with resources expenditures needed to be corrected. PPBS became the solution to this problem. The early RAND report on the PPBS defines the terms as follows:

> Planning is an analytical activity carried out to aid in the selection of an organization's objectives and then to examine courses of action that could be taken in the pursuit of objectives. . . . Programming is the function that converts plans into a specification schedule for the organization. . . . Budgeting is the activity concerned with the preparation and justification of the organization's annual budget. . . . Operations consist of the actual carrying out of the organization's program. . . . Evaluation is the function that evaluates the worthiness of programs.[4]

While the system is highly formulaic, having such structure is imperative for conducting purposeful R&D that is connected to operational requirements. Without such a system, R&D could be conducted with little or no linkage to use cases in an operational environment.

Should this occur, systems could be developed that would never be fielded by the services or employed in the execution of operational plans.

The PPBS established a system that linked strategies, plans, operations, R&D, system acquisition, and life cycle support. The PPBS had also been nested within the DoD, where each of the departments, agencies, and geographic and functional commanders had a role in the process. Each was required to prepare their own PPBS documents. Strategic documents that established requirements for the joint force were developed annually, planning was revised frequently to respond to changes in the operational environment, programs (to include the R&D) that supported the DoD's myriad requirements were established, and resources were committed.

This early management and integration system has continued to see growth and adaptation. The PPBS, now called the PPBES to account for managing execution of the system, employs large numbers of people working in the Pentagon and across the senior commands who are actively participating in this strategy-to-resources process. *Joint Vision 2010*—a conceptual approach to future warfighting requirements that was written in 1996—led to a significant part of this growth, serving as the intellectual firepower behind the development of the Joint Requirements Oversight Council, which is used in developing joint military capabilities, identifying operational gaps, developing recommendations for filling these gaps, recommending resource expenditures consistent with the priorities and required capabilities, and establishing performance requirements for measuring the forces' ability to achieve operational outcomes.[5]

By the late 1990s, the PPBES saw further growth in complexity and organizational requirements with the development of the Joint Capabilities Integration and Development System. One of the goals of the JCIDS process was to reinforce the connected model of capability development (and by extension R&D) to ensure that the capabilities developed would support the warfighting requirements of the combatant commanders.[6]

While the PPBES began in the DoD, most departments and agencies in the US government now have a similar method for strategy to resource funding and the development of capabilities. It is fair to say, however, that the DoD model has a complexity and size that dwarfs all others. It is within the overarching PPBES that focused technology development is intended to occur.[7] This implies developing and justifying programs that have a waiting customer looking for the particular technology, capability, or system under consideration.

The discussion of connected and focused technology development should be placed in context. While addressing operational concerns—especially within the departments and agencies—has been elevated to great importance (and prominence), there remains a recognition that pure science, discovery, and basic research must continue to have a place within the R&D. The NSF and the internal research and development (IRAD) discussed previously are designed to provide space for these types of efforts.

However, another feature of the PPBES, which many departments and agencies either follow directly or have components of, implies that R&D competes directly against operational requirements for resources. In this battle, operational requirements normally prevail, as the here and now trumps the longer-term R&D efforts. Some might argue this priority is shortsighted, yet for those in R&D, this must be factored into the execution of research and development programs. Those programs with an interested customer who is willing to allocate the resources to the technology development and eventually to acquisition programs are far more likely to survive funding cuts.

As an example, when I first arrived in the S&T Directorate in 2011, the organization had just absorbed a 56 percent reduction in funding. Our approach to gaining restoration of the R&D budget was to more closely align with the department's operational components, which had the greatest need for technology and the largest budgets. As a result of this alignment, the directorate's annual budget more than doubled over the next two years to over $1 billion.

While PPBES represents a structured approach to linking strategy to resources, it is not without its critics. Some have criticized PPBES as being overly complex and inflexible. Recent efforts such as the DIUx (now DIU, which removes the *experimental* from the title) were designed to bring more innovation to the Pentagon. Some have called for PPBES reforms, arguing that the detail required in running the PPBES does not result in better use of DoD resources. Adaptations to allow greater flexibility will likely occur, but the fundamentals of the PPBES are also likely to remain in place.

ESTABLISHING THE OPERATIONAL NEED

A natural tension exists between the operator and the technologist. The operator thinks about today's problems and looks for ready solutions to allow her to do the job more effectively, efficiently, and safely. The technologist thinks about solving tomorrow's problems, looking at how technology could provide important solutions. To complicate matters, the operator has a deep understanding of the context within and the constraints under which her duties are performed. The technologist, on the other hand, may lack this perspective and might have only a limited understanding of the actual requirements.

In thinking about this problem, two cartoons come to mind. The first describes a group of people in blindfolds standing around an elephant, each touching a different part of the animal. One is touching the trunk, another the tail, and some are touching the legs and ears. In other words, each person is describing the elephant, but from a different perspective. The caption reads, "OK, now go ahead and describe the elephant." With such vastly different viewpoints, actually gaining a consensus of what an elephant is becomes impossible. This represents the operational view, in which operators have very different perspectives depending on where they fit in the system.

The second cartoon is of a guy—we will call him a technologist—alternately looking at a small opening in a machine and over his shoulder at a large contraption that is clearly five times the size of the opening. The caption for this cartoon is, "So you want me to fit this in here?"

The technologist, having little understanding of the operational environment and constraints, has developed a technology that is intended to become part of a larger technology and, ultimately, a larger system.

Neither of these cartoons portends a happy outcome. The operator cannot adequately describe the operational context, and the technologist has little operational understanding of how the technology is to be employed.

Bridging this divide requires closing the knowledge gap between the operator and technologist to find solutions that lead to the purposeful application of capabilities (including scientific knowledge or the collection of techniques, methods, or processes) for practical purposes. Developing operationally relevant technologies is also essential to crossing the valley of death. When a technology is developed that fulfills an operational requirement, the potential for resulting in a fielded system greatly increases.

Industry has recognized that operators and R&D specialists might speak slightly different languages and therefore could benefit from a translator to help them communicate effectively. Some companies have developed these translator functions, having either a tech-savvy operator or an operationally savvy technologist interface with the intended end user and the R&D specialists. Companies such as Oracle have employed this function to ensure that the technology developers continue to work on operator priorities. At first glance, one might see these translator teams as nothing more than business development; however, that would be diminishing their importance. Technology companies expend significant resources on R&D and want to ensure they are spending the technology development resources wisely. The technology translators serve as one mechanism for doing so.

Another important way to bridge the divide between operators and technologists is through industry days and trade association meetings that allow technology developers to demonstrate their capabilities. The most effective of these industry days are those held in operational environments where a more casual mixing between the tribes—operators and technologists—can take place.

During one such event when I was in DHS S&T, an operational demonstration took place at Camp Roberts in California featuring unmanned aerial systems and command and control software. The three-day event allowed for viewing and testing the gear in an operational environment. The technologists could hear firsthand about operational problems, gauge how well their systems could address those concerns, and even make changes to their systems (some in real time) to better reflect what they heard to be the operational issue. The operators also benefited greatly from having the opportunity to understand the current state of the technology and its limits and perhaps even to adjust expectations.

Another such event took place at Texas A&M in November 2015. The event featured Homeland Security technologies deployed in realistic scenarios. The event was cosponsored by General Dynamics's EDGE Network, which consists of more than 550 industry, university, and nonprofit partners, and Texas A&M Engineering Extension Services. Through this exchange, industry gained valuable insights into operational requirements and the challenges faced by the government. While these requirements were not linked directly to upcoming acquisitions, such interactions help to focus corporate IRAD and provide better insights into the government's needs. Academia benefits as well, as colleges and universities can provide insights on how to focus their research portfolios. This academia-supplied R&D has become a major source of basic and applied research and early-stage development for the government. Government benefits, as representatives can gain key insights into the state of the latest technologies at relatively low cost in a learning and nonthreatening environment. This also provides a basis for a more informed customer with a better understanding of the state of technologies in key areas.

Industry associations such as the Aerospace Industries Association, the Association of the United States Army, and IEEE host periodic meetings that allow for a similar mixing of government and industry. Most provide opportunities for vendors to display products and conduct presentations on their technologies. These industry associations also provide a forum for businesses to collaborate to solve problems, promote standards, and develop solutions for industry-related issues.

Challenge competitions are focused demonstrations in which participants are required to complete a set of defined tasks. Typically, the government announces a challenge for a technology to meet a set of operational specifications, allows industry to prepare their technologies, and hosts a competition for interested parties. The prize is normally a modest dollar award that may not even cover costs but allows the winner to claim bragging rights. One such challenge was sponsored by DARPA in 2015 and inspired by the 2011 Fukushima nuclear disaster in Japan. An earthquake and tsunami had caused major system failures at the Fukushima Daiichi Nuclear Power Plant. The result was a buildup of explosive gas, several explosions and fires involving the nuclear material, and the release of a large radioactive plume. Conditions at the plant limited human intervention and made prolonged human exposure dangerous.[8]

The goal of the challenge was to design an autonomous system that could operate in a radioactive environment. The challenge called for the teams' offerings to complete eight tasks within an hour: (1) driving a utility vehicle, (2) exiting a vehicle, (3) opening a door, (4) drilling a hole in a wall, (5) turning a valve, (6) completing a surprise task, (7) walking over a pile of rubble or clearing a path through debris, and (8) walking up a short flight of stairs. The competition required the robots to complete the competition without a tether and while having only intermittent communications. Twenty-five teams participated in the DARPA Robotics Challenge Finals. A team from South Korea took home the $2 million first prize. Two teams from the United States took second and third, winning $1 million and $500,000, respectively.[9]

Two relatively small federal programs are designed to foster innovation and promote small businesses, the Small Business Innovation Research (SBIR) and Small Business Technology Transfer (STTR) programs. The SBIR program encourages domestic small businesses to engage in federal R&D that has the potential for commercialization. Competitive awards allow small businesses to conduct focused R&D in topic areas of interest to the government. Twelve departments and agencies in the federal government have SBIR programs, and each has the flexibility to administer their programs based on their needs and the guidelines established by Congress. The SBIR program has three phases tied to the R&D life cycle: (1) establishing the technical merit, feasibility, and commercial potential of the proposed R&D, which is an award of less than $150,000, (2) continuing R&D efforts, not to exceed $1 million, and (3) pursuing commercialization.[10]

STTR also provides funding for innovative R&D through an expansion of the public/private sector partnership to include the joint venture opportunities for small businesses and nonprofit research institutions. On STTR, the website claims its "most important role is to bridge the gap between performance of basic science and commercialization of resulting innovations."[11] Both the SBIR and STTR require award recipients to continue to move through technology maturity gates to continue to receive funding for their programs. Five departments and agencies participate in this program; it also has three phases that are the same as the SBIR program. The goals are to stimulate technological innovation, foster technology transfer through cooperative R&D between small businesses and research institutions, and increase private sector commercialization of innovations derived from federal R&D.[12]

Broad Agency Announcements (BAA) allow the federal government to solicit R&D proposals in focused topics. The program targets basic and applied research and early-stage development. As with the SBIR and STTR programs, the departments and agencies have significant flexibility in the topics that are selected, the award amounts awarded, and the overall administering of the program.[13]

A cooperative R&D agreement (CRADA) is yet another mechanism for developing and formalizing a relationship between government and industry for R&D efforts. CRADAs can be established at any point in the R&D life cycle from basic and applied research to technology development. The earlier in the technology development life cycle a CRADA is established, the more the R&D resource burden will be shared between the government and industry.[14] At DHS, we employed a CRADA for the development of a vaccine for foot and mouth disease.

In this case, the government had done the research and early development, and the CRADA was designed to mature the vaccine through the later stages of technology development, ensuring commercialization, and crossing the valley of death. The program was successful, and the vaccine is now commercially available to consumers.

To more rapidly develop operational capabilities and introduce flexibility into government procurement systems, several departments and agencies—including DoD, DoE, HHS, DHS, Department of Transportation, and NASA—have been authorized to use Other Transaction Authority (OTA) for certain prototype, research, and production projects. In effect, the use of OTA provides DARPA-like capabilities to other government organizations, helping to streamline procedures that previously could constrain typical contracts, grants, and cooperative agreements. OTAs can be used for industry partners and academia, and are intended to foster partnerships with non-traditional partners.

This section has discussed programs that have been established to improve the connectedness of R&D between industry and government customers. Despite numerous formal programs and informal opportunities for encouraging communication between operators and technologists, focused R&D remains challenging and requires constant attention to ensure that purposeful technology is developed that is likely to become operationally useful.

Before we leave this section, IRAD should be discussed in more detail. Earlier, it was briefly mentioned, but few specifics were presented. IRAD refers to internally sponsored R&D work that is considered strategic. That is, laboratory funding is put toward science or technology development that may not have a sponsor but may have a longer-term benefit. The funding for IRAD is generated through the equivalent of a "tax" that is levied on funded projects. The amount of IRAD varies, but somewhere in the neighborhood of 5 percent is considered reasonable by many labs. IRAD can include a range of programs from basic research to later-stage developmental activities, and the award of these funds is normally at the prerogative of the laboratory director or an IRAD community at an individual research facility.

METHODS OF TECHNOLOGY DEVELOPMENT

No attempt will be made to exhaustively list and describe the various technology development tools that are available; however, a brief review of some of the more widely used tools will be beneficial later for our discussion on assessing technology. Tools used in project management can provide foundations for tracking science and technology, research and development, and innovation and transformation.

Classic program management methods such as the critical path method (CPM) can be used to employ a structured way to think about technology development. By defining the tasks that need to be accomplished, the subordinate technologies that need to be integrated, and the sequence and time frames for accomplishing each task, one is able to establish a "critical path" through the development cycle.[15]

While using CPM for the development of a single technology is possible, some may find it more useful when looking at development of a larger system. In this way, the component tasks, integration requirements, delivery of materials, and system tests could be sequenced and tracked to ensure system completion in a timely manner.

Related to the critical path method conceptually is the earned value management system (EVMS), which allows for identifying the tasks required to complete a project (in our case, a technology development project) and developing cost, schedule, and performance measures for each. Once objective measures have been developed for each task, one can develop a plot over time that compares cost, schedule, and performance to determine if a project is progressing as planned. Any deviations can be examined and corrective actions taken as necessary.[16]

Technology road maps provide a related, less formal approach to technology development. A technology road map provides the sequencing of the different requirements inherent in an R&D project. The road map can be a graphic depiction or a logical listing of tasks, relations and dependencies between them, and sequencing. Of course, by adding the elements of cost, schedule, and performance, a technology road map could have the same rigor as CPM or EVMS.

A related tool employed in the DoD is the use of Operational Views (OVs) in capability planning. The OV-1 is the highest level and provides a graphical depiction of the system under consideration. It describes a mission, class of mission, or scenario; the main operational concepts and interesting or unique aspects of operations; and the interactions between the subject architecture and its environment and between the architecture and external systems. The purposes of the OV documents are to map out the organizations, operational concepts, process flows, relationships between organizations, and activities that govern the various elements involved in a key operational area.[17] The various OV documents DoD uses are listed in appendix D.[18]

Some may find the notion of using structured methods to do R&D a bit unrealistic, as technology development does not necessarily proceed in a linear manner and may include high-risk elements that may cause the technology to have to return to an earlier TRL. Even if methods such as CPM, EVMS, technology road maps, and OVs are not fully applied, however, having a structured way to consider technology development can be useful.

Road maps are not just used in DoD any longer. For example, the Biomedical Advanced Research and Development Authority (BARDA), within the Office of the Assistant Secretary for Preparedness and Response in the US Department of Health and Human Services, uses such an approach in managing the vaccines, drugs, therapies, and diagnostic tools for public health emergencies that are in its pipeline.

The Intelligence Community also has an end-to-end process that looks to align the R&D programs in progress with operational issues that must be addressed. Such collaboration serves to ensure that key areas of R&D are covered and that resources are being applied as necessary to develop operational solutions. Recently, the IC has made a significant effort to look toward open-source technologies that can be adapted for their purposes; an example includes the use of social media exploitation.

In DHS S&T, road maps and a portfolio management approach were employed to ensure that focused R&D was being conducted. These efforts were intended to assess the types of activities being done and determine whether the portfolio and the individual programs in areas such as explosives detection, screening and vetting, border security, chemical and biological defense, cybersecurity, disaster preparedness and response, and first responder technologies were likely to result in programs with operational relevance.

ASSESSING AN R&D PORTFOLIO

During my time in the Science and Technology Directorate of DHS, we employed a portfolio review process that sought to ensure the R&D that was conducted and funded through our programs contained specific attributes that increased the likelihood of a successful transition to operators. This same framework (with some minor modifications) was quite familiar to me, as it was also used to review L3 Corporation's R&D portfolio during my time with industry.[19]

The R&D portfolio analysis framework consisted of four overarching areas: operational focus, innovation, partnerships, and project quality. Each of the areas was further broken down into elements that provided a structured way to evaluate the utility of the R&D. Appendix E lists the assessment areas and the subordinate elements used in conducting such a portfolio review.

Examining an organization's R&D portfolio on a periodic basis provides important opportunities to assess and examine the individual projects and overall portfolio. Other benefits of conducting such a review are to give the organization's senior leaders an appreciation for the full range of R&D being conducted and the funding profiles of the individual R&D programs and to provide an opportunity to reshape the portfolio should it be determined that restructuring is in order.

The four major areas assessed during our review process provided indications of the priorities of the leadership. Operational focus ensured that the projects were aligned with operational needs. They might fit directly into a customer's current concept of operation, or perhaps they might fill a critical operational gap and therefore be likely to transition to use once development is complete. In some cases, these needs could reflect near-term requirements, while in others, a longer horizon for the R&D was considered. One of the keys in a portfolio review was to understand the relative weight of the collection of projects. Were a majority weighted toward satisfying near-term requirements, or were they weighted toward examining needs well into the future? Neither of these cases was desirable. Rather, the goal was to have a healthy balance between the near- and longer-term R&D programs. Operational focus also considered the degree to which the customer was invested in the technology development, whether in the form of funding, in-kind support (e.g., people to test the technology), interest in the effort, or even demonstrated commitment to transitioning and procuring a technology after development.

Innovation pertained to the approaches and the likely result, attempting to ascertain whether the R&D being conducted was likely to result in unique outcomes or whether others in the community were already working on the development of this technology. It also focused on whether a new operational capability or efficiency was likely to be developed as an output of this R&D program. Innovation in this assessment also sought to ascertain whether the technology development was likely to be feasible or whether it would rely on other technologies that had not yet reached maturity.

Partnerships allowed the leadership to assess the degree of cooperation and collaboration in a project. Partnerships were also useful for sharing the costs and risks associated with conducting R&D and could be mutually beneficial in cases where each partner brought unique capabilities and experiences to the effort. Partnerships also sought to leverage the body of knowledge that came before, identifying what R&D has been conducted and whether it would be suitable for the envisioned operational problem the organization needs to address. Even if others had worked in the space, an R&D project might still be required to address the specific operational problem in question. However, by looking at the body of work that has been done previously, one would begin by being better informed and perhaps even further along in the R&D process. Perhaps through the work of others, the R&D could begin with development rather than having to invest in costly applied research.

Project quality examined the programmatic elements of the R&D effort, looking at the project management and whether the R&D project was likely to transition to operational use. Project clarity, cost, and timelines were considered, as they directly pertain to the project management, while transition likelihood examined whether the R&D was likely to transition into an acquisition or operational program.

In the case of the portfolio review conducted by the leadership of DHS S&T, a premium was placed on maintaining a delicate balance between achieving near-term results and undertaking longer-term, more risky projects that had the potential for larger payoffs. There was no percent goal across the entirety of the portfolio, and in fact, when one looked across the portfolio, the balance varied greatly. In areas such as first responder technologies, which relied on scouring relevant areas of development to find capabilities that could be readily absorbed to fill operational gaps, the focus was on near-term solutions and development of standards to drive technology solutions. For biodefense, riskier R&D was being undertaken in niche areas such as sensing and biodiagnostics, two areas where we assessed that breakthroughs were required, and we hoped to at least stimulate that community.

CHAPTER 8

MANAGING TECHNOLOGY IN AN UNMANAGEABLE WORLD

U nderstanding the technology life cycle and characterizing it through technology readiness levels (or TRLs) has provided a useful framework for considering technology development. While TRLs allow one to assess the maturity of the technology, they do not seek to capture explicitly when a technology will become widely available to the public. Nor does the TRL framework identify when a technology is likely to be used in dangerous and unintended ways. In the absence of such considerations, history demonstrates that technologies will continue to proliferate largely unfettered until governments, stakeholders, and the public become aware of the unintended consequences and potentially dangerous outcomes. Armed with this awareness, the reaction is normally to attempt to "manage" technology by developing control measures and limits in the form of policies, regulations, and laws.

The very concept of managing technology implies being able to change the trajectory of a scientific field or individual technology. Yet as history has repeatedly demonstrated, technology development relates more to human preferences and needs than to a rigid plan to either promote or retard technologies. In fact, our definition of technology— the purposeful application of capabilities (including scientific knowledge or the collection of techniques, methods, or processes) for practical purposes—provides insights into whether one is likely to succeed in directly guiding a field of study. We should expect technology to develop based first on response to a perceived need, second on whether the basic science and principles are understood and resources applied, and third on any constraints or limiters in place (related to either the feasibility of the science or human constraints).

This chapter will discuss the ways humankind has sought to control the proliferation of technology, primarily for security or commercial purposes. While competition for technology and scientific discovery are not new, taking a structured approach to controlling or managing technology is. Appendix F provides examples of various control mechanisms that have been applied throughout history.

We will also see evidence that these efforts have met with varying degrees of success. Such outcomes are less related to the control measures employed and more a result of the nature of the technology in question. In cases where controlling technology has been successful, the complexity of the technology, limited availability of components, and/or resources required were key factors. In those cases where technology could not be successfully managed, there were low barriers to entry and/or strong motivations for technology proliferation and use.

CONCEPTUAL UNDERPINNINGS

Dangerous uses of technology, technological surprises, and the erosion of the US dominance across many key technologies have signaled a need to consider different approaches to the placing of limits, controls, and management oversight on key technologies. The inadequacy of current methods for controlling technology has become part of an increasingly public debate.

The rapidly increasing development pace and growing availability of technology combine to create a tipping point where monitoring—or, in some cases, limiting activities—and proliferation—of a specific technology will prove prudent. Such control measures could include voluntary compliance, codes of ethics, institutional review boards, policies, regulations, laws, and international agreements. However, implementing control measures—or attempts to manage technology—could also have deleterious effects by limiting the development of an important technology that has potentially great societal benefits.

The question of technology limits must also be considered within the context of an increasingly globalized R&D environment. In such an environment, limits on technology development would invariably be implemented differently across the globe. Therefore, any attempts to limit technology development and proliferation must be considered in the broader context of balancing technological advancement, global perspectives, and the safety and security of society.

History indicates that placing limits on technologies often occurs only once they are identified as national security imperatives. For more than a century, arms control has centered around placing limits and even prohibiting activities surrounding conventional, nuclear, chemical, and biological weapons. Export controls have long sought to limit the proliferation of various technologies that were key enablers of commercial technology yet had national security implications, calling these technologies dual use. More recently, beginning in 2009, the Obama administration conducted a broad-based review to assess the current utility of the export control system and make changes to account for the evolving global environment. The review sought to focus the export control system on fewer items while placing greater restrictions on those key areas that should be protected. The process involved an interagency effort, led by National Security Council staff.

However, these controls have largely been developed in reaction to observed behaviors rather than being forward-looking strategies. Historically, no repeatable, methodical approach for identifying when technology limits or controls should be put in place has been developed. The field has suffered from a lack of rigor and inability to predict when technologies are likely to become readily available and could become dual-use risks. US national decisions on technology proliferation largely reflect the subjective, decentralized prerogatives of the responsible government departments and agencies that own the processes and programs in particular technology areas. The examples below highlight the reactive nature of technology control mechanisms:

- Drones were being used by early adopters and resulted in several near collisions with commercial aircraft before the Federal Aviation Administration began to regulate their use.
- Dual-use research of concern (DURC) (as discussed in chapter 3), sponsored by the government, funded two performers to assess whether the H5N1 influenza virus could be made more transmissible while still retaining its virulence. While the studies were conducted with appropriate control, the biodefense community underwent important self-reflection, resulting in the development of a presidential executive order governing DURC.
- The Apple Corporation–Federal Bureau of Investigation standoff resulting from the 2015 terrorist attack in San Bernadino, California, and the government's request to have the perpetrator's cell phone opened to aid in the investigation brought privacy expectations and encryption concerns to light long after cell phones had become commonplace.
- Genetic data collected from individuals desiring to gain insight into their backgrounds had been used without their knowledge for solving crimes and identifying previously unknown relatives. While in the case of crime, some may consider this acceptable (i.e., the ends justify the means), these cases highlight the loss of privacy and unintended consequences of this technology.

A CASE STUDY: THE INABILITY TO "CONTROL" TECHNOLOGY

Humans have long sought to control their environment. However, controlling technology is no easy task. Too much control of early technologies could stifle the science, knowledge, and innovation likely to result in important advances. Too little control of a technology could allow dangerous uses with potentially catastrophic outcomes. To this choice, some might say, "Pick your poison."

The goal should be to find a path that satisfies the need for both discovery and prudence. Yet as humankind has witnessed repeatedly, technology normally develops before control mechanisms are in place. It is only after the technology has reached a certain point, when society perceives its potential, that control mechanisms are considered.

Social media provides an interesting case in this regard.[1] Many point to 1997 and the establishment of a website called Six Degrees as the birth of social media. The website allowed users to share personal information. It lasted for only four years. Despite the demise of Six Degrees, the internet continued to expand and increase the number of users, which would serve as the perfect environment for the social media revolution that was to come.

Facebook and Twitter came next. In 2004, Mark Zuckerberg launched Facebook, which seemingly shifted overnight the amount of personal information that we would be willing to share with each other and the world. Two years later, when Twitter was launched, it would provide a unique platform for sharing news, providing personal updates, and sharing one's innermost thoughts in 140 characters or fewer.

Use of social media continues to proliferate. Anyone with a connected device likely has a social media account; many even have multiple accounts. "News" and updates are generated at an alarming rate on these services. A person can spend the entire day doing little more than immersing himself in social media information. But just like other technologies, social media is dual use, as we have come to see over the twenty-plus years it has been in existence. One need only consider the government-sponsored Russian influence operations surrounding the 2016 presidential election to see evidence of this dual-use activity.

What is equally clear is that the developers of social media and those responsible for leading these companies had not previously considered the potential for their platforms to be used as attack vectors for misinformation. Initially some of the executives of these social media companies even pushed back on the idea that they could be targeted for misdeeds and were providing a platform, perhaps even a breeding ground, for misinformation. In fact, since these services have been in existence, we have seen evidence of the dark side of social media, including online bullying, misinformation, and scams. And as the

2016 US election also demonstrated, these social platforms could have serious national security implications as well, in this case allowing a nation-state to use social media to attack another state.

During congressional testimony in October 2017, lawyers from Facebook, Google, and Twitter testified that the Russian influence campaign surrounding the 2016 presidential election had been much more pervasive than they originally had believed possible. More than 126 million Facebook users had viewed propaganda provided by Russian operatives. Twitter identified almost 2,800 accounts controlled by Russians and more than 36,000 bots that had tweeted 1.4 million times during the election. Google identified more than 1,100 videos with forty-three hours of content provided by Russian sources.[2]

The February 2018 indictment of thirteen Russians by Special Counsel Robert S. Mueller III provides unmistakable evidence of the trail of mischief and considerable damage that has been perpetrated through social media platforms. Russian efforts to manipulate the American people during the 2016 election and American society in general have become clear. The investigation identified continued efforts to manipulate US elections and further erode America's confidence in its institutions.[3]

As with other technologies that we have discussed, control mechanisms for social media are only being developed in the aftermath of the discovery of misuse or abuse of a technology. However, one need only look at the internet—originally developed as an information-sharing platform, with virtually no security features in its earliest incarnation—to understand that such misuse was entirely possible and part of a broader issue concerning Information Age technologies. After all, it took the misuse of these technologies to develop the concept of cybersecurity.

Social media standards for protecting data and limiting the detrimental uses of these platforms had begun to be considered seriously only after revelations of election influence during the 2016 election. Earlier allegations of cyberbullying (some even resulting in suicides) and reselling of social media data did little to turn the concerns into concrete actions. Furthermore, many of the changes that are underway in social media have been forced upon the companies in the interest of self-preservation rather than being undertaken through altruistic efforts to improve services and protect consumers.

PROTECTING TECHNOLOGY

This section considers how to think about protecting technology, both for economic purposes and from those individuals and nations that might seek to use technology for malicious or dual-use ends. For the latter, it considers mechanisms established for protecting technology from people and for protecting people from technology.

Five categories of activities for protecting technology have been identified: (1) international laws and treaties (including arms control), (2) international bodies such as the World Health Organization, (3) coalitions of the willing, (4) US unilateral actions, and (5) national laws and best practices. Example activities for each category are provided in appendix F. In considering these categories, several notes are in order.

Dividing these activities into five discrete categories as we have done could leave the misimpression that they are distinct. In fact, quite the opposite is true. For example, national laws, best practices, and US unilateral actions are related to, and in some cases derivative of, international laws and treaties. The same can be said of international bodies such as the World Health Organization, which has developed and enforces requirements for public health reporting that are embedded in national laws, policies, regulations, and best practices.

The examples presented are certainly not the only activities being undertaken in these categories; rather, they are intended to be representative of the types of activities employed for managing technologies. Each will be discussed in greater detail within this section, and we will conclude by considering how successful the United States has been in managing or controlling proliferation by relying on such tools.

Given how rapidly technology has proliferated over the last hundred years, understanding the mechanisms that are in use and their effectiveness could be beneficial in looking to the future and perhaps even developing new mechanisms to limit the spread of potentially dangerous technologies.

Before delving into this topic and by way of a disclaimer, the prospects for establishing control mechanisms to prevent all misuses of technology would be an unrealistic goal that could surely not be achieved. Therefore, developing reasonable policies and programs that reduce the chances of misuse of technology, lengthen the time between the development of a dual-use technology and its prospects for widespread illicit use, and protect economic prerogatives may be determined to be the best possible outcome for many technologies.

In such an approach, the delicate balance between serving the interest of greater technological advancement with the safety and security of society must be considered at all times. To this end, a risk-based framework for guiding efforts to assess when additional scrutiny or even control measures could be required becomes essential. At all costs, a "one size fits all" or heavy-handed approach to technology management should be avoided, as it could have a stifling effect on technology development yet do little to avoid the concerns resulting from the misuse of technology.

NATIONAL LAWS AND BEST PRACTICES

National laws, policies, and regulations (including intellectual property rights and patents) provide legal and regulatory mechanisms for managing and controlling technology. Best practices represent activities such as standards of conduct, norms of behavior, and codes of ethics designed to assist in managing technology development and use.

By the definition of what it means to be a state, nations are responsible for controlling activities that occur within their sovereign borders. National laws and best practices, which can be voluntary or mandatory in the case of regulations and policies, are the mechanisms by which states exercise this control. States failing to exercise such control could be considered ungoverned or loosely governed spaces. In such cases, international mechanisms could be applied to force a nation to comply with international law.

While national laws are based on a country's prerogatives, culture, and norms, they should conform to the international laws in specific technology areas. As a result, in the case of genetically modified organisms (GMOs), one might see different approaches in the United States, China, and the European Union. So long as each comports with international law on the issue, this should be deemed legal. To further elaborate on this example, GMOs are grown in twenty-eight nations around the world but are prohibited or banned from cultivation in more than three dozen countries. The European Union has taken a hard line on the issue, while China and the United States only restrict certain types of GMOs. In the case of GMOs, international conventions for marking products have been developed. Still, assuring that GMOs are not inadvertently introduced into a non-GMO country is difficult.[4]

Weapons of mass destruction (WMD) provide another example. National laws regarding WMD provide a basis for national (as well as state and local) prosecution for those engaged in this type of illicit activity. Such offenses are covered in 18 US Code § 2332a—Use of weapons of mass destruction, which pertains to anyone who "uses, threatens, or attempts or conspires to use" a WMD. The statute covers a wide range of activities both here in the United States and for US citizens regardless of the location of the infraction.[5] Of note, the national laws on WMD are completely aligned with and complementary to international law on the subject, which is covered by several relevant arms control treaties.

Policies and regulations are also frequently established to govern activities in certain technology areas. In some cases, violations of policies and regulations can be cause for legal prosecution, whereas in other cases, perhaps no legal remedy exists. For example, if a dangerous experiment was conducted in a material science lab, but no law has been violated, perhaps the only remedy is disciplining or firing the employee.

However, if the employee in the materials lab was found to have been printing 3-D guns that were operational, the activity could be illegal and controlled under laws regarding illegal possession of a firearm.

Community initiatives such as institutional committees can also serve to limit certain prohibited or dangerous R&D activities and behaviors. For example, institutional bio committees have been established at laboratories where life sciences work is conducted. Prior to an experiment being conducted, the committee considers the work and either approves, disapproves, or requires changes to the work plan prior to the initiation of the R&D. Other technology areas have different mechanisms to ensure the appropriate use of the science, technology, and resources. For example, prior to conducting an experiment employing an explosive detonation chamber, one R&D laboratory required review and approval of the procedures to be used in the research.

Many fields of study have well-established codes of ethics that govern norms, behaviors, and activities. The Hippocratic Oath— *primum non nocere*, or "first, do no harm"—reflects a simple, yet clear and unambiguous code of ethics governing medical treatment.[6]

In fact, most professional fields whose work impacts the public have an ethical code of conduct. Plumbers, construction workers, and law enforcement all have ethical standards to which a professional in the field must adhere. The IEEE code of ethics has a provision in its code specifically for dealing with the use of related technology: "to hold paramount the safety, health, and welfare of the public, to strive to comply with ethical design and sustainable development practices, and to disclose promptly factors that might endanger the public or the environment."[7] In the future, it seems almost no area of R&D will be impervious to the call for a code of ethics. Many are asking whether software developers should also be governed by such a code.[8]

Intellectual property (IP) rights provide a mechanism for protecting R&D and guarding trade secrets. IP provides for the "assignment of property rights through patents, copyrights and trademarks."[9] Assuming the R&D has been appropriately documented and a patent has been submitted—either a provisional or full patent—a monopoly is established, and the IP is protected. This has important ramifications for commercial protections for a technology in question. For example, if a technology has not been protected through a patent, the developer of the technology could lose access to the right to use the technology.

While the World Trade Organization (WTO) recognizes IP, and most nations have agreed to and ratified the WTO-crafted agreements in their parliaments, national laws have nuances, meaning that the application of IP across the world is not uniform.[10] Even between trading partners with significant dealings, IP issues can be contentious. Consider the relationship between the United States and China, in which China gains illicit IP from three sources: "hacking, spying and requiring companies to form 'joint ventures' with local companies, which then have access to sensitive information."[11]

While hacking and spying can be thought of as part of the normal—even if distasteful—competition between the nations, the use of joint ventures to create an institutionalized loss of IP is problematic. For those caught engaging in hacking or spying, prosecutions would be possible, as in the 2014 case of five Chinese military hackers charged with cyberespionage for targeting US corporations and a labor organization for commercial advantage. A news release on the indictments alleges the hackers conducted "computer hacking, economic espionage and other offenses directed at six American victims in the US nuclear power, metals and solar products industries."[12]

The theft associated with the joint ventures, however, does not lend itself as readily to legal remedies and is an inherent part of companies that have joined together for a common and limited economic purpose.

The IP issue is made more complex by differences in culture between the United States and China, as China places greater emphasis on common good in the settlement of IP issues and is therefore more interested in the community-wide good that comes from dissemination and less interested in the individual benefit that is a higher priority for the United States.[13]

OTHER US INITIATIVES

The United States has several other initiatives designed to protect US R&D. Four that will be discussed include export controls, focused non-proliferation efforts, border security and customs, and the Committee on Foreign Investment in the United States (CFIUS).

The US Departments of State, Commerce, Homeland Security, Treasury, Defense, and Energy have critical roles in export control activities within the United States and outside its borders. Concerning export control, State has responsibility for commercial exports of defense articles and defense services. These responsibilities are exercised through the Directorate of Defense Trade Controls. The centerpiece of these activities is the US Munitions List (USML), which consists of twenty-one categories of hardware, technical data, and defense services to be protected.[14] Specifications for each category are available in the International Traffic in Arms Regulations (ITAR). The exercise of this authority is designed to restrict and control military and defense technologies to protect national security objectives.[15]

Commerce has a complementary role to State, which is exercised through the Export Administration Regulation. The EAR uses the Commerce Control List (CCL) to control dual-use items that have a primary commercial or civil application but also have the potential for military or weapons applications. The CCL is composed of ten categories[16] and five groups that describe the type of technology being considered.[17]

Under the Obama administration, an Export Control Reform Initiative was undertaken. Many in industry had complained that export control regulations placed an undue burden on them and resulted in a loss of revenue in some cases. The goal of the reform initiative was to restructure export control categories and build "higher walls around fewer items."

The DHS role in export control is on the enforcement side, which is exercised primarily through Customs and Border Protection (CBP) and Immigration and Customs Enforcement (ICE). CBP operates at our borders to prevent threats from entering the United States. CBP also runs the National Targeting Center in Sterling, Virginia, which looks to identify and detain travelers and detect cargo that could threaten the United States, as well as to facilitate the legitimate flow of trade and travel.[18] Initiatives include the Container Security Initiative (CSI) and Customs-Trade Partnership Against Terrorism (C-TPAT).[19] CSI seeks to identify and inspect at foreign ports any maritime container that could be used to deliver a weapon. C-TPAT achieves faster throughput and greater cargo security through its voluntary public-private partnership that increases regulatory compliance standards and licensing.

ICE has responsibilities for customs enforcement through its Homeland Security Investigations (HSI) organization, which operates the Export Enforcement Coordination Center. CBP is responsible for exercising authorities across a number of missions that relate to R&D, including import safety, IP rights issues, and trade agreements. The goal of the center is to "detect, prevent, disrupt, investigate and prosecute violations of US export control laws."[20] The DHS responsibilities are to prevent unlawful technology transfers from the United States to another party and to prevent illicit goods including technologies from entering the United States.

Treasury and Energy have more narrowly defined roles related to proliferation of weapons of mass destruction. In the case of Treasury, authorities include placing controls on transactions and freezing assets.[21] Energy, through the National Nuclear Security Administration, focuses on control of nuclear technology.[22]

Nonproliferation activities are also part of US and global initiatives to control the flow of potentially dangerous dual-use technologies, focusing on technologies that could support illicit WMD programs. An example is the Nunn-Lugar Cooperative Threat Reduction (CTR) program developed at the end of the Cold War to assist in securing dangerous material and dismantling WMD capabilities. The effort looked to eliminate WMD facilities; safeguard and destroy dangerous material; build facilities overseas in partner nations, such as epidemiological labs that support public health programs; and increase cooperation with former officials and technicians involved in the former Soviet weapons programs.[23] The CTR program, which was established through the passage of the Soviet Threat Reduction Act in November 1991, has been extended beyond the former Soviet Union nations, primarily to developing countries in Asia and Africa, to assist in building technical capabilities for developing biodefense capacity, enhancing border security, and establishing institutions in partner nations.

CFIUS operates to assess whether certain business transactions could adversely affect national security. The program determines whether foreign investment or sale of a company or assets to a foreign entity will be detrimental to US national security. A Congressional Research Report lists the five investments that have been blocked through the CFIUS mechanism:

1. In 1990, President Bush directed the China National Aero-Technology Import and Export Corporation (CATIC) to divest its acquisition of MAMCO Manufacturing.
2. In 2012, President Obama directed the Ralls Corporation to divest itself of an Oregon wind farm project.
3. In 2016, President Obama blocked the Chinese firm Fujian Grand Chip Investment Fund from acquiring Aixtron, a German-based semiconductor firm with US assets.
4. In 2017, President Trump blocked the acquisition of Lattice Semiconductor Corp. of Portland, OR, for $1.3 billion by Canyon Bridge Capital Partners, a Chinese investment firm.
5. In 2018, President Trump blocked the acquisition of semiconductor chip maker Qualcomm by Singapore based Broadcom for $117 billion.[24]

While CFIUS has not been used frequently since being established by executive order of President Ford in 1975, it represents an important, and perhaps increasingly relevant, mechanism for protecting US technology.

INTERNATIONAL INSTITUTIONS AND COALITIONS OF THE WILLING

This category of international collaboration involves multilateral efforts to control proliferation. Forums such as the Missile Technology Control Regime (MTCR), Nuclear Suppliers Group, Australia Group, and Wassenaar Arrangement are composed of like-minded nations that seek to cooperate to control dual-use technologies. The MTCR limits the proliferation of missiles with a payload of five hundred kilograms or more and a range in excess of three hundred kilometers.[25] The Nuclear Suppliers Group seeks to reduce proliferation of dual-use nuclear technologies through the development and use of guidelines.[26] The Australia Group looks to harmonize export control for technologies that could contribute to the development of chemical or biological weapons.[27] Finally, the Wassenaar Arrangement seeks to gain greater transparency and responsibility "in transfers of conventional arms and dual-use goods and technologies."[28]

Other international efforts provide mechanisms for countering proliferation of WMD capabilities. Examples include the Proliferation Security Initiative (PSI)[29] and Global Initiative to Combat Nuclear Terrorism (GICNT). PSI provides increased international cooperation in interdicting shipments of WMD, their delivery systems, and related materials. Specifically, PSI assists partner nations in training for interdiction of transfers of suspected WMD capabilities, supports in strengthening national laws and promoting information exchange, and seeks to reduce proliferation by providing training support for combating illicit networks. GICNT is designed to "prevent, detect, and respond to nuclear terrorism."[30]

Organizations such as the World Health Organization (WHO), Food and Agriculture Organization of the United Nations (FAO), and World Organization for Animal Health (OIE) are helping to protect populations from emerging infectious disease, feed them, and improve their quality of life. The exchange of scientific and technical capacity underpins the efforts of all of these organizations.[31]

INTERNATIONAL TREATIES AND LAWS

International treaties and laws seek to limit certain behaviors, activities, and weapons types through the development of legally binding mechanisms. Arms control has been used to limit military activities and preparations both prior to and during war and stems from what's known as just war theory, which limits activities and behaviors to prevent unnecessary human suffering.[32]

Treaties can be either bilateral or multilateral. Such agreements normally have procedures for assessing compliance called verification protocols and consultative forums that allow for arbitrating any infractions or disputes that may arise. The language of arms control is quite precise and legalistic, calling for parties to an agreement to adhere to specific requirements, provide data on activities and weapons holdings, and even submit to announced and surprise inspections.

The US history of arms control goes back to the 1817 Rush-Bagot Treaty between the United States and United Kingdom, which demilitarized the Great Lakes and Lake Champlain region of North America. Later, in 1871, the United States and United Kingdom signed the Treaty of Washington, which led to further demilitarization and dispute resolution and, through arbitration, resulted in a payment by the United Kingdom and an accompanying return of friendly-nation status.

Since the 1899 Hague Convention, which placed limits on deployment of forces and new armaments, the international community's use of arms control has accelerated with all manner of international treaties, from ones that place limits on military deployments and activities to those banning certain types and classes of weapons and other dual-use technology.[33] Such legally binding mechanisms have also been used to establish norms of behavior and codes of conduct for warfare and the use of military technologies.

Some arms control agreements place limits on or outright ban certain types of weapons. The Biological Weapons Convention (BWC) eliminated an entire class of weapons, requiring states to refrain from any activity to "develop, produce, stockpile or otherwise acquire or retain microbial or other biological agents or toxins . . . that have no justification for prophylactic, protective or other peaceful purposes."[34] The Chemical Weapons Convention (CWC) banned the use of chemical weapons and toxic industrial chemicals on the battlefield. The Convention on Cluster Munitions sought to eliminate (or at least limit) the use of these dangerous munitions.

Several arms control agreements have focused on nuclear weapons exclusively. These treaties have halted proliferation activities, eliminated destabilizing weapons, and reduced weapons holdings. The Nuclear Nonproliferation Treaty (NPT) promotes disarmament, nonproliferation, and peaceful uses of nuclear energy. Stockpile reductions have been negotiated through treaties such as the Strategic Arms Reduction Treaties (START), with the latest being the New START, which would limit US and Russian holdings of strategic nuclear weapons to 1,550 delivery systems. Some treaties, such as the Fissile Material Cutoff Treaty (FMCT), have sought to halt production of dangerous radiological material for use in nuclear weapons, in a sense attempting to control activities related to the nuclear fuel cycle; others, such as the Comprehensive Test Ban Treaty, have attempted to limit testing involving nuclear weapons and subcomponents of nuclear weapons. The Intermediate Nuclear Force Treaty eliminated a class of weapons, including ground-launched ballistic and cruise missiles with a range of 500 to 5,500 kilometers.[35]

Treaties have also been established to control activities in specific geographic areas. The Outer Space Treaty and Moon Treaty provided limits on activities in space. Regional nuclear-weapon-free zones prohibit deployment and use of nuclear weapons and, in some cases, have been interpreted as a ban on ships transiting the area that have nuclear power propulsion systems.

Other international treaties have resulted in limits not related to arms control, such as the Kyoto Protocol to reduce greenhouse emissions and the Rome Agreement, which established the International Criminal Court for bringing to justice those individuals suspected of crimes against humanity.[36]

Legally binding frameworks such as the 1994 Agreed Framework between the United States of America and the Democratic People's Republic of Korea (DPRK) have been established to eliminate nuclear weapons capabilities on the Korean Peninsula. This mechanism was not successful, as the DPRK failed to live up to the terms of the agreement.

This discussion was not intended to be a complete elaboration of all treaties and legal instruments that are in place or being negotiated. Rather, it was intended to provide an overview of how international agreements can serve to limit technologies and related activities considered dangerous to humankind or our environment.

EFFECTIVENESS OF TECHNOLOGY CONTROL REGIMES AND EFFORTS

In the earlier section "Protecting Technology," a range of activities was provided that included international laws and treaties (including arms control), international bodies (such as the WHO), coalitions of the willing (such as PSI), US initiatives (such as CTR), and best practices (such as standards and norms of behavior). But do these activities work to prevent proliferation of dual-use technologies? Unfortunately, the history is mixed on this question.

In considering this issue, one needs to examine the effects of the increasing connectedness and global trade inherent in globalization, which make enforcing limits on illicit or undesirable behaviors and activities more difficult. To this end, a more complete accounting of the history of globalization will be useful.

The World Trade Organization describes what author G. John Ikenberry calls the "first age of globalization" as follows:

The early 19th century marked a major turning point for world trade. Although the outlines of a world economy were already evident in the 17th and 18th centuries—as advances in ship design and navigation led to Europe's discovery of the Americas, the opening up of new routes to Asia around Africa, and Magellan's circumnavigation of the globe—it was the arrival of the industrial revolution in the early 1800s which triggered the massive expansion of trade, capital and technology flows, the explosion of migration and communications, and the "shrinking" of the world economy, that is now referred to as "the first age of globalization."[37]

Going back to the nineteenth century and the first age of globalization may seem irrelevant when considering today's rapidly expanding democratization and proliferation of technology, but they are assuredly inextricably linked. Before the nineteenth century, trade was, as characterized by one account, "persistently low."[38] Beginning in the nineteenth century, with advances in technology for shipbuilding, navigation, and even agriculture and manufacturing (which generated goods in excess of what people required for subsistence), trade increased 3 percent annually until about 1913.

World Wars I and II resulted in a significant reduction in global trade and a reversal of the earlier trends. With the end of the wars, trade began to increase until it reached a plateau in the 1950s coinciding with the beginning of the Cold War.

Looking specifically at the period of 1980–2011, exports of goods and commercial services increased from $2.31 trillion to $22.27 trillion. This change of an order of magnitude over a thirty-year period is impressive, but it does not tell the entire story. From 1980–1985, exports actually declined slightly. After the end of the Cold War in 1990, the global export markets went on a tear, with exponential increases over the twenty-year period from 1990–2011.[39]

Trends in trade have also been fueled by lowering of supply chain costs and an increase in preferential trade agreements. Since the 1930s, these trends have also been fueled by a lowering of costs for sea freight and passenger air transport.[40] Looking at trade agreements shows a slow increase from 1950–1977 to fifty agreements, holding steady with only modest increases from 1977–1990, an exponential rise to more than 280 agreements by 2000, and then a leveling off until 2010 at the end of the data set.[41]

All this is to say that through globalization and the related trade, more people, goods, and services are flowing throughout the world, developing networks and pathways that support licit trade and travel. On the other hand, these same networks and pathways also provide a means to ship illicit goods, including dual-use technologies, across the globe.

Since the end of World War II, the United States (at times even without our allies' support) has sought to limit sales of technology. After the war, it was discovered that 85 percent of the aircraft parts and machine tools used by the Mitsubishi Aircraft Company came from the United States.[42] Formal mechanisms were established to limit the flow of goods to certain nations. The Marshall Plan, better known as the European Recovery Program, prohibited any recipient European country from exporting to a nonparticipating European country (meaning a communist country) any product that might contain a US-supplied commodity that would ordinarily be refused a US export license."[43]

These export controls were specifically designed to limit a potential adversary's ability to exploit US technology. They were frequently used to good effect during the Cold War. Computer technology provides an example of their effectiveness: The Soviet bloc understood the importance of computers for future military applications, both as standalone systems and as components for other technologies, such as for fire control, missiles, and satellites. However, the Soviet bloc lacked the supplies of components, trained computer specialists, and factories to manufacture these systems. One assessment indicated they were four years behind the West in the development of computer capabilities. The United States was able to retain this "strategic advantage" through an embargo on the shipping of computers to the East.[44]

After the Cold War ended, a deliberate decision by the United States and our allies called for easing restrictions on high-technology goods. The Coordinating Committee for Multilateral Export Controls cut in half the number of high-tech items that could not be shipped to Russia and other former Warsaw Pact nations. Some military-grade equipment remained tightly controlled. One account describes the protected technology as including night-vision systems and supercomputers that were considered essential to maintain the West's superiority in military technology over the Soviet Union. Restrictions also continue on high-performance fiber optics, some computer software systems, thermal imaging equipment, advanced machine tools, advanced semiconductor-manufacturing equipment and some super-strong carbon fibers and composites.[45]

In the previous computer example, the embargo was aided by the minimal trade between the two blocs resulting from East-West tensions during the Cold War. However, today the environment has shifted considerably. Global supply chains crisscross the world, and few countries remain isolated. Even the DPRK, which has been subjected to significant United Nations sanctions, has found ways to circumvent them, thus limiting their overall effectiveness.

For example, the DPRK relied heavily on reverse engineering for its rocket designs; support from Egypt in the 1970s; likely, additional support from the Soviet Union and then Russia in the 1980s and 1990s;[46] and other external sources of technology over more than forty-five years. More on the DPRK example will be discussed later in the chapter.

Given the nature of technology, nonauthorized dual-use technologies are routinely shipped around the globe, as the DPRK missile example demonstrates. Furthermore, many of the same technologies used in common items such as cell phones have applicability for defense and security proposes. The Obama administration's Export Control Reform Initiative was intended to account for these changes by placing "higher fences around fewer items," meaning that priorities only needed to be in place to guard against technologies judged to be of the greatest concern.

Despite export controls to limit the proliferation of military equipment and dual-use technology, knowledge products, and training for personnel, the record of nonproliferation activities has been mixed. While the term *proliferation* often refers to WMD capabilities, for our purposes, we will take a more expansive view. Simply put, evidence suggests that nonproliferation and counterproliferation can slow but do not stop the flow of critical military and dual-use technologies. Several cases highlight this assertion.

Examining the proliferation of nuclear technologies provides an interesting point of departure for answering this question. Today, there are nine nations that have or are believed to have nuclear weapons. They include the NPT-designated nuclear weapon states (China, France, Russia, United Kingdom, United States) and other states with declared nuclear weapons (India, North Korea, Pakistan). One other state is believed to possess nuclear weapons (Israel), and four other states—former Soviet states Belarus, Kazakhstan, and Ukraine, as well as South Africa—have given up their nuclear capabilities. Iran's status remains uncertain given the US decision to withdraw from the Joint Comprehensive Plan of Action (JCPOA) nuclear agreement. Other states including Libya, Syria, and Myanmar investigated and began development of a nuclear weapons capability, but these programs are no longer active.

Only the five original states had indigenously developed weapons, implying the other states relied on assistance from other nations (or individuals) though proliferation pathways. Examination of several nuclear nonproliferation cases is instructive for understanding the issues: Israel's secret nuclear path, India's special arrangement, Pakistan's A. Q. Khan network, and the collapse of the North Korea framework. Each provides interesting implications for current and future nonproliferation regimes.

Israel's nuclear weapons program began in the 1950s based on what can be considered a strategic imperative, the survival of Israel. The French secretly agreed to assist Israel by helping them build a nuclear reactor and underground reprocessing plant to produce plutonium.[47] The development of a nuclear weapon capability required twenty tons of "heavy water," which neither the French nor the United States would supply. The Israelis turned to Norway, which agreed to sell the "heavy water," completing Israel's nuclear pathway.[48]

India's development of nuclear weapons did not follow a straight development path. Early statements on nuclear weapons saw India as pushing toward a strong nonproliferation regime, calling for the halt to all nuclear weapons testing in 1954. India pursued a peaceful nuclear program throughout the 1950s and early 1960s. By 1965, India began to reconsider its stance given the rivals in the region, Pakistan and China. India had also developed the complete nuclear fuel cycle and later conducted its first nuclear test in May 1974. Along the way, India was aided by the United States, which supplied a small amount of plutonium for research in 1966.[49] India also used plutonium from a Canadian-supplied research reactor that contained US heavy water for conducting its first nuclear explosive test, and received indirect aid through information from Western scientific journals.[50]

Pakistan's A. Q. Khan was at the center of the world's largest nuclear technology proliferation network, having developed nuclear weapons for Pakistan and subsequently illegally provided nuclear weapons technology to Iran, Libya, and North Korea. Khan earned a doctorate in engineering and took a position at a uranium-processing plant in the Netherlands. During his time in the Netherlands, he was "copying designs for centrifuges and compiling a list of potential companies that could provide Pakistan with the technology to produce highly enriched uranium for nuclear weapons."[51]

Despite his activities having attracted the attention of Dutch law enforcement, he was not arrested and continued his efforts. Upon returning to Pakistan, Khan became essential to the development of an indigenous uranium enrichment process using the stolen centrifuge designs. When the United States became aware of Pakistan's nuclear efforts, President Jimmy Carter imposed economic sanctions, attempted to block loans to Pakistan from the World Bank, and pressured nations (including France) to halt sales of nuclear technologies. With the Soviets in Afghanistan and the United States needing a base of operations in close proximity, the United States placed support to Afghanistan ahead of nonproliferation policy. By then, Pakistan was well on its way to developing nuclear weapons, but Carter's lifting of

sanctions and $400 million in economic and military aid likely short-
ened the timelines for developing a nuclear capability.[52]

North Korea was aided in the development of a nuclear capability
by A. Q. Khan. Based on this assistance, North Korea developed nuclear
production facilities in the 1990s. Through the 1994 Agreed Framework, North Korea agreed to provide a complete history of its nuclear
activities, mothball the nuclear production facilities, and refrain from
reprocessing its spent nuclear fuel rods. However, simultaneously, North
Korea began a covert reprocessing effort. Despite repeated overtures
and direct pressure to cause North Korea to halt its nuclear weapons
program, continued testing occurred in 2006, 2013, 2016, and 2017.
North Korea's last test was thought to be a hydrogen weapon, estimated by some to have a yield of up to 250 kilotons.[53]

North Korea's missile development also demonstrates the difficulties with halting the flow of these types of technology. It is likely
much of the missile technology development has been indigenous, with
heavy reliance on "reverse engineering," as several of the designs were
reminiscent of Soviet systems. North Korea also procured entire rocket
systems, for example from Egypt in the 1970s. It is also possible that in
the 1980 and 1990s, North Korea received technical support from the
Soviet Union and then Russia.

Regardless of the exact sources of technology, what is clear is that
North Korea has been able to flout the stiff sanctions that have been
imposed. A 2013 UN panel established to monitor the implementation of sanctions found evidence of noncompliance including impor-
tation of fine-grain graphite, classified documents, and machine tools
for missile technology development. After one launch, recovered rocket
debris was discovered to have components sourced from China, Switzer-
land, the United Kingdom, and the United States, in addition to Scud
missile components and 1980s Soviet parts. Of note, all these nations
are members of the Missile Technology Control Regime except China.[54]

What is not known is whether nations knowingly provided the
technology or it was covertly obtained. Regardless, nations have a
legal obligation under the 2004 UN Security Council Resolution 1540

to control exports to prevent the proliferation of dangerous missile, nuclear, and military technologies.

A 2017 United Nations report details how North Korea uses "technology smugglers and financial institutions" to develop missiles and nuclear weapons. In characterizing North Korea's efforts, the report states,

> The Democratic People's Republic of Korea is flouting sanctions through trade in prohibited goods, with evasion techniques that are increasing in scale, scope and sophistication. The Panel investigated new interdictions, one of which highlighted the country's ability to manufacture and trade in sophisticated and lucrative military technologies using overseas networks.[55]

The North Korea example highlights that despite international efforts, a strong sanctions regime, and direct engagement at high levels bilaterally and as part of the six-party talks, the DPRK was able to develop WMD capabilities.

Chinese theft of intellectual property demonstrates another dangerous technology proliferation window. The theft of IP—not just in military technology—has been sanctioned and, in some cases, undertaken through active participation by government organizations and employees. The IP loss has been in a wide range of technology sectors such as automobiles, aviation, chemicals, consumer electronics, electronic trading, industrial software, biotech, and pharmaceuticals. One source noted that IP theft costs the United States more than $600 million annually, with most of the loss coming from China.[56] Another reported theft of more than $1 trillion in IP on the F-22 and F-35 aircraft allowed the Chinese to close the technology (and capability) gap with the United States on fighter aircraft capabilities.

A report from November 2017 highlights that the US Department of Justice had issued indictments on three Chinese nationals for IP theft. They were accused of stealing from several companies, including Trimble, a maker of navigation systems; Siemens, a German technology company with major operations in the United States; and Moody's Analytics.[57]

Anecdotal evidence suggests that proliferation of technology continues at an alarming pace. Many of the incidents can be traced to not having adequate safeguards in place. In others, decisions have been made to prioritize a political issue over technology proliferation and IP protection. International regimes and export controls have proven leaky and unable to stem the flows of technology across the world.

Turning to the question of what fields or technologies are likely to be either good or bad candidates for control through formal or informal mechanisms, examples can be useful.

Let's start with computers. Earlier, we described how the control of computers was important to blocking the Soviets from gaining access to these capabilities and their potential second- and third-order uses, either as components or subcomponents of other technologies or systems. This made sense as the technology was early and had not proliferated. However, now project to the present and consider whether control of computers makes sense in the future. We would certainly not be able to place limits on the personal computer industry, given the maturity of the technology and how widespread the use has become. However, one could easily foresee that in the field of high-performance computing[58]—an emerging field likely to be extremely important for applications regarding massively parallel problems and large data sets—limits on technology export could still be applicable and useful.

The use of personal computers provides another interesting question to consider. Will it be possible to control the proliferation of knowledge in the computer industry and the actions of individuals? Obviously, laws can be used to bring people to justice for infractions perpetrated while using a personal computer. However, with such a low cost of entry (measured in a couple hundred dollars for the cost of a PC) and the democratization of knowledge, which in some cases has allowed even "untrained" people to operate with significant effect, managing the technology is far more difficult.

Consider the case of a seventeen-year-old who received top honors in an US Air Force–sponsored hacking competition for finding 30 valid vulnerabilities. The field included 272 security researchers who discovered a total of 207 vulnerabilities.[59] While the individual was obviously quite skilled and the label "untrained" does not exactly pertain, his skills did not require state capabilities to develop.

Fields with the highest cost of or barriers to entry have the greatest potential for being monitored and controlled. Nuclear technologies, which require specialized processed materials, fall into this category. The cost of entry includes access to raw material, processing of the material, a nuclear reactor, and if one is developing a nuclear weapon, access to enrichment capabilities. The range of systems and technologies to be mastered also creates barriers to entry. Each step along the path requires mastery of the technology. The footprint required means it would be difficult to hide, thus potentially increasing the probability of detection of the illicit activity.

What about 3-D printing? Today, 3-D printing is in the initial stages of broad adoption. Still, there are barriers to use. The techniques have not been demonstrated across a broad range of materials. The strength and failure properties still must be better understood. There is also a significant start-up cost associated with buying a high-end, high-quality 3-D printer. However, with continued development and maturity of the technology, many of the barriers are likely to be lowered. Still, there may be opportunities to protect some cutting-edge 3-D printing capabilities, even if only for a modest period.

CONCLUSIONS

In many areas, advances are so rapid that they are outpacing policy makers' ability to place adequate controls on S&T, R&D, and I&T with potentially devastating consequences—and certainly before treaties, laws, policies, and regulations are in place.

Proliferation of dual-use technologies is allowing individuals to have statelike capabilities. Areas of science and technology that once were the exclusive domain of nations, such as Global Positioning Systems, have become part of our daily lives. Multinational corporations are complicating placing controls on science and technology proliferation.

Laws and treaties, other international forums such as the Wassenaar Arrangement and the World Health Organization, coalitions of the willing, and unilateral actions—in combinations and at times individually—have historically been used to shape outcomes, influence national and international actions, and enhance US security. With the proliferation of science and technology across the globe, the effectiveness of many of these institutions is being called into question.

Global corporations, supply chains, and knowledge proliferation will continue to make protecting key technologies more challenging. In looking to the future, the United States should look for other opportunities to more effectively "manage" technologies while protecting competitive advantages and US economic and security interests.

PART III

THE FUTURE

I fear the day that technology will surpass our human interaction. The world will have a generation of idiots.

—Albert Einstein

If we continue to develop our technology without wisdom or prudence, our servant may prove to be our executioner.

—Omar Bradley (General, US Army)

CHAPTER 9

THINKING ORTHOGONALLY

Often, highly structured thinking and analysis can be quite constraining and limit considerations of other possibilities or potential outcomes. This can be especially true in areas with considerable uncertainty and seemingly limitless opportunities.

In this chapter, we will propose the need for becoming more expansive in our thinking about technology. When considering the future of technology development, linear and inflexible thinking can give one a kind of tunnel vision in which only a single future can be envisioned, thus limiting the ability to sense the opportunities and challenges on the horizon. Thinking orthogonally can serve to open the aperture more broadly and allow for consideration of a range of possible futures.

The term *orthogonal* means "at a right angle to" or "very different or unrelated; sharply divergent." Thus, an orthogonal thinker is one who approaches problems in different or nontraditional ways.

So where does orthogonal thinking come from? First, having people with a diversity of talents, experiences, and worldviews allows for more robust thinking. Such thinking is less likely to be monolithic or consider only a single or perhaps traditional point of view. Diversity is essential in developing technologies with appeal, acceptance, and cultural sensitivity toward a broader cross-section of society, and it also includes having people from multiple technology specialty areas collaborating on problems. Consider if a group of people with common experiences and educational backgrounds is asked to solve a problem. Such a group is highly likely to develop traditional consensus approaches when confronted with an issue for consideration. But having a diversity of specialties could provide opportunities for different approaches and concepts to be identified.

Second, different groupings of personalities can introduce nonlinear or orthogonally oriented thinking into the solution of a problem. To illustrate this, consider whether you are a waffle person or a spaghetti person. A waffle person likes to put things into the square holes of a waffle, much like filling in a matrix. Such a person is likely to be very linear and organized. Thoughts will likely move in a linear manner, covering goals, objectives, subordinate objectives, and elements. The spaghetti person is likely a nonlinear thinker more comfortable dealing with uncertainty and ill-defined issues. Bringing together waffle and spaghetti people can provide interesting insights when contemplating a complex issue.

Third, finding people with different risk tolerances can be very important to finding orthogonal solutions for addressing difficult problems. If a monolithic approach is employed, the math can be rather straightforward in the calculation of risk. However, if different people with different risk tolerances are asked to assess risk, they are likely to arrive at very different conclusions and highlight outcomes—both good and bad—that had not been considered before.

Fourth, structured approaches that cause people to engage to address a common problem can expose orthogonal thinking. The use of role-playing scenarios or the Delphi method can cause individuals and groups to interact and expose such varied thinking. These sorts of tools are likely to expose a wider variety of approaches to problem-solving.

History contains numerous examples of orthogonal thinking. The Manhattan Project, discussed previously, provides one such example. Consider the multitude of experiences and fields represented within the staff who developed the atomic bomb. General Eisenhower's leadership during the D-Day invasion likewise demonstrates orthogonal thinking, employing what many thought was a riskier invasion point along the French coast and gaining strategic surprise, rather than selecting the less challenging, yet more predictable, point of attack. At a very technical level, saving the Apollo 13 crew after the capsule explosion in space required the use of orthogonal approaches in overcoming myriad technical challenges in developing life-support systems, conducting trajectory analysis, overcoming power shortages, and returning the crew safely to Earth.[1]

While most projects could benefit from orthogonal points of view at times, large, complex multidisciplinary efforts have the potential to see the most benefit. The greater number of potential paths and the greater the complexity of a project, the greater the potential benefit one is likely to see from orthogonal thinking. Translating this to technology development, the same principle holds true. Larger, highly complex systems with components themselves composed of highly complex systems can also benefit greatly by orthogonal thinking.

Some might take issue with orthogonal thinking being highlighted as a new concept, arguing that it is nothing more than possessing critical thinking skills. I would counter that what is unique is the deliberate introduction of orthogonal thinking and thinkers into complex technology problems to more fully explore the range of possible outcomes and develop solutions.

While orthogonal thinking comes from a variety of unrelated disciplines and perspectives, it can provide novel insights into complex problems. Brian Arthur's use of evolutionary biology to describe the evolution of technology provides such an example. During the remainder of the chapter, examples of thinking orthogonally for some technologically difficult problems will be presented. A mix of historical examples and future areas that would benefit from orthogonal thinking are discussed.

PACKET SWITCHING AND DIGITAL COMMUNICATIONS

Modern data communications are based on orthogonal thinking that occurred in the 1960s and has become the basis for all digital communications throughout the world today. Prior to the invention of packet switching, communications relied on individual circuits to carry message traffic. Each individual circuit was dedicated to a customer and therefore was normally idle unless that customer was either sending or receiving communications.

Packet switching provided a means for using the available communication streams more efficiently. It relied on a series of protocols and methods for decomposing individual communications into smaller partitions called packets, where each packet had a header and payload. The header provided the address location, and the payload contained the message traffic. Packet switching relied on another important feature to operate: the path traveled by the message packets and the order in which they arrived at the address location were irrelevant. When they arrived, the packets were reassembled into the same format as the original message from the sender.

Decomposing and recomposing messages established a virtual circuit at the precise point in time at which it was required to get the communication through. Standardized connectionless protocols, such as Ethernet, Internet Protocol (IP), and User Datagram Protocol (UDP), and connection-oriented protocols, such as X.25, frame relay, Multiprotocol Label Switching (MPLS), and Transmission Control Protocol (TCP), provided the technical controls for packet switching.[2] Despite latency resulting from different packet paths and the assembly and reassembly of the messages, the entire process is done instantaneously and therefore not obvious to the sender or receiver of the message.

While certainly not obvious to users of the internet today, the development of digital communications and packet switching has a lengthy history involving numerous fields of science and technology. Built in the 1950s for providing assured communications in the event of a nuclear first strike by the Soviet Union, it relied on communications technology, network theory, and queueing theory.

These concepts came together in ARPANET, developed by the US Advanced Research Projects Agency (ARPA) in the late 1960s.[3] ARPANET was originally intended for connecting four university computers; however, the development of TCP/IP protocols in the 1970s made it possible to expand the network to become the network of networks that we call the internet.

While the internet of today is far more complex than the early ARPANET, the foundations of digital communications, including packet switching, remain its backbone. Orthogonal thinking can be seen in several aspects of this example. The "outside the box thinking" that provided the basis for packet switching and digital communications began with the development of a survivable communications network. The technology was developed and refined through experimentation involving experts in several related fields. The concept of breaking a message into multiple component parts for transmission and reassembling it at the destination represents both orthogonal thinking and a counterintuitive approach for delivering messages rapidly and accurately.

IDENTITY MANAGEMENT

Identity management applies to a broad range of activities designed to establish a person's identity. It is an "organizational process for identifying, authenticating and authorizing individuals or groups of people" for granting access.[4]

Such access can be granted for a physical location, information, or a computer network, to name a few. Access is normally granted based on an established set of criteria. For example, if one is attending a sporting event, requirements normally include holding a ticket and undergoing screening at the gate to ensure that prohibited items are not brought into the stadium. For computer networks, network administrators and security personnel require that users are screened, have appropriate permissions to access the network, and have valid passwords to secure their accounts and training to understand their requirements as users.

The Department of Homeland Security's Transportation Security Administration (TSA) has undertaken an orthogonal approach to identity management for ensuring aviation security. In 2017, the TSA screened more than 770 million passengers through 440 federalized airports in the United States. The volume of passengers screened annually calls for a taking a risk-based approach to screening and vetting.

A risk-based approach combines a mix of screening equipment, sensors, information, real-time communications, and processes designed to achieve aviation security while minimizing passenger disruption. It is based on a simple yet orthogonal concept—that people should be screened according to the likely risk they present to the aviation system. Using this approach, TSA has increased the flow of passenger screening while maintaining the security of the system.

A risk-based approach to screening begins from the understanding of what constitutes the greatest threats and risks to the aviation system. The manifests of aircraft are prescreened at the National Targeting Center run by Customs and Border Protection before aircraft can depart from overseas locations. Passengers considered low-risk receive lower levels of screening as they process through airports. Those enrolled in the US Global Entry program who have voluntarily provided personal information in advance receive a different level of intrusiveness in the physical screening, going through magnetometers rather than the advanced imaging technology that uses automated target-recognition software in screening passengers. Frequent travelers with established flying profiles will likely be considered lower-risk as well, even if they do not have the Global Entry credentials—passengers participating in frequent-flier programs through the airlines would also fall into this category. Those passengers not enrolled in the Global Entry program undergo a standard set of screening protocols.

However, those passengers who have been identified as presenting an elevated risk receive more intense primary screening and perhaps even secondary screening. Passengers on aircraft originating from certain destinations could also receive increased scrutiny. The systems have been designed to allow TSA agents to more carefully scrutinize suspicious passengers or passengers who do not have known aviation profiles.

Without TSA's risk-based screening and vetting approach, the number of agents, equipment, and facilities required for screening would have continued to grow. This orthogonal approach to identity management demonstrates how technology—really, the collection of technologies represented by the mix of screening equipment, sensors, information, real-time communications, and processes—have been combined into a comprehensive, risk-based security system.

INFRASTRUCTURE REVITALIZATION AND SMART CITIES

Infrastructure revitalization and establishment of smart cities could each benefit by orthogonal thinking. The number of stakeholders involved in both endeavors means that many governmental, industry, science, and technology areas as well as private citizens will be involved in developing and promoting solutions. The voices of these stakeholders will need to be heard and the best ideas incorporated into final designs to build effective and efficient solutions.

In considering infrastructure revitalization, it is important to recall how US cities, transportation networks, and industries evolved. Early settlers to North America established communities along the eastern coast of what is now the United States. Through exploration and hope for expanded economic opportunities, the move west began. Waterways provided a source of ready transportation for this expansion. Others traveled by wagon, sometimes stopping their journey when confronted by a river or a mountain range. This helps to explain the patterns of settlement we see today.

As communities were established, industries began to grow. Factories were built along waterways to facilitate the movement of raw materials and finished products. Shops were established to support the communities. Power plants were built to provide electricity to the factories. Electrical lines were run to provide power to the factories and eventually the rest of society. When railroads were built, they often connected major cities and the transportation hubs that supported them. In other cases, the railroads cut long, straight paths across the country, and cities were established to support railroad operations.

If this description seems rather haphazard, it is—for good reason. Cities and the supporting activities were based on practical considerations at the time they were established. Few, if any, could be considered planned communities. Even where urban planning was done, the urban sprawl has taken over. Consider Washington, DC, which was laid out in a grid. Today, the population of DC is just over 600,000, with another 3.5 million or more who live in five adjacent suburbs.[5] If one were to design the city today, it would undoubtedly look different.

The same could be said of communities across the United States. We would certainly want to adjust population centers, transportation networks, and business locations to better suit today's needs. We would also want to factor risks and best practices into redesigning our nation's cities. Consider how large populations cluster in some of the most risk-prone areas. While coastal areas comprise only 20 percent of the land, they account for more than 50 percent of the population, with 39 percent of the US population living in counties subject to significant coastal flooding during flood events, which have a 1 percent chance each year of occurring. Six of the ten most expensive US natural disasters were caused by coastal storms.[6] If we were redesigning America's cities, undoubtedly changes would be made.

Such inefficiencies are certainly problematic, and they are made more severe through aging and obsolete infrastructure designed to support America in a different era with the technologies of days past. Today's electrical grid was largely developed in the early twentieth century with highly inefficient alternating current (AC) transmission lines that were state-of-the-art when they were installed but would be considerably more efficient if direct current (DC) electricity lines had been used. Unfortunately, modern technologies such as high-capacity battery storage and superconducting cables were not available when the original lines were installed. As with other types of infrastructure, the locations of the large electricity-generating plants and substations are based on historical need rather than on the usage requirements of today, which in many cases have changed dramatically.

To further exacerbate the infrastructure issue, several layers of utilities have been appliquéd on top of our cities: water, electricity, sewer, natural gas, traffic, telephone, cellular telephone, and distribution centers and hubs. Virtually all have aged and require maintenance and replacement. Some have become obsolete—for example, landline telephone networks that have been replaced by cellular communications.

Our nations' bridges, railway system, and roadways require significant attention as well. The highway system was established and built beginning in 1953 during the Eisenhower administration. Many of the bridges date from the earliest days of the highway system and lack the weight and volume capacities to support today's traffic. One estimate puts the annual cost of maintaining highways and bridges through 2030 at $65.3 billion per year.[7] By another estimate, the total cost to repair aging US infrastructure is more than $1 trillion, and that does not include the cost to modernize our systems.[8]

Our railroad infrastructure continues to age as well. Washington DC's Metro system is suffering from decades of underfunding and thirty years of deferred maintenance. While other nations have invested in modern railways, the US railway system has not been upgraded to modern capabilities. The lack of positive train control, which has resulted in several near misses, train crashes, and derailments highlights this lack of modernization.

There are certainly many potential claimants for infrastructure revitalization efforts, but determining which projects should receive priority must be done. In an ominous warning, one expert identifies the major question as "Where to wisely invest those dollars so that we don't piss away $1 trillion on solving problems that don't exist or in areas that won't get you the biggest return?"[9]

This is a good point to reintroduce the need for orthogonal thinking. Repairing or revitalizing yesterday's infrastructure would be a self-defeating strategy. Building replacement power plants with the same or increased capacity may be solving yesterday's problem. Today, opportunities exist for augmenting centralized power generation locations with microgrids that can serve as decentralized power distribution networks using local power generation in the form of renewables (e.g., solar, wind, geothermal). Such a strategy could reduce the inherent risks associated with a single large facility becoming nonoperational through an accident, failure, or attack. Regardless of the source of the outage, resilience could be enhanced through the development of a more decentralized power generation and distribution system.

Thinking orthogonally would also cause one to question whether autonomous vehicle possibilities could affect driving habits, vehicle ownership, requirements for roads and highways, and parking in and around cities and places of work. As autonomous vehicles enter our society, we should expect corresponding changes to occur. What should we expect the effect of autonomous vehicles to be on the long-haul trucking industry? In all likelihood, the use of autonomous aerial vehicles or drones for local delivery of goods could also be a disruptive change that would need to be factored into estimates for the capacity of our roads in the future.

Another area that seems ripe for technological innovation and orthogonal thinking is the internet. While the internet began as an information-sharing platform, over the past two decades, it has evolved into a combination of a ubiquitous communications network and a utility. The cellular network and mobile communications represent the ubiquitous presence, while the internet connections in homes and businesses represent the portion of the internet that functions like a utility. Businesses, schools, and governments are all requiring access to the internet, and those without high-speed connections are disadvantaged. The Internet of Things will only exacerbate this problem. Perhaps now is the time to consider the internet of the future and determine how the current internet will need to evolve to satisfy future requirements.

Up to this point, we have been discussing orthogonal thinking as applied to revitalizing infrastructure. But opportunities also exist in designing smart cities, either through new development or inserting technology into existing cities.

One description of a smart city states,

> Smart city initiatives involve three components: information and communication technologies (ICTs) that generate and aggregate data; analytical tools which convert that data into usable information; and

organizational structures that encourage collaboration, innovation, and the application of that information to solve public problems.[10]

While the development of purpose-built smart cities has been identified as a possibility, in reality, most smart-city applications will be embedded in existing cities. Some foreign nations have plans for development of smart cities built from scratch, but this is the exception. For example, Saudi Arabia is planning to build Neom, a $500 billion city designed to explore new economic opportunities. Neom will be an independent zone with its own regulations and social norms so as to attract top talent in trade, innovation, and creativity.[11]

For our purposes, we will concentrate the remainder of the discussion on the application of smart-city technologies in current US cities. Several US cities have already begun to incorporate such technologies.[12] City leaders combined with technologists have combined to monitor, measure, and control various aspects of the city's daily life, from garbage collection to the structural integrity of the city's infrastructure.

One can certainly envision that as the IoT network continues to grow, the possibilities for integrating more smart functions and capabilities will also grow. Over time, it is likely that cities will confront difficult and complex questions concerning the integration of these new technologies. For example, one would expect that drones will be pervasive, either for delivery of goods, communications, transportation, or observation. How will their flight paths be deconflicted? Will virtual highways in the sky be developed for this purpose? And who is responsible should a delivery drone fall out of the sky and injure someone? And as smart-city technologies continue to proliferate, what will be the effect on individual privacy?

Such complex questions would undoubtedly benefit from an orthogonal perspective. Just because a technology can be developed or fielded does not mean it is beneficial or worth the cost. These sorts of decisions will require integration and coordination by government leaders, industry, technologists, and concerned citizens working in unison on these complex issues.

CANCER MOONSHOT

President Obama launched the Cancer Moonshot Initiative shortly after the death of Beau Biden, the son of Vice President Joe Biden and his wife, Jill, from brain cancer in May 2015. The goal of the program was to make more therapies available to more patients and improve our ability to prevent cancer and detect it at an early stage.

The initiative brought together a wide range of people from the medical, R&D, and industrial communities to find solutions to the challenges. With the authorization of the 21st Century Cures Act, Congress allocated $1.8 billion over seven years. By necessity, the effort needed to gain a better understanding of cancer at all levels, from the science to the treatment of patients.

Opportunities for R&D are anticipated to be expanded under the Cancer Moonshot Initiative (since renamed Cancer Breakthroughs 2020). Laboratory centers have been established for the development of deep tumor and immune profiling that would assist in cataloging various types of tumors. Identifying and validating biomarkers will be necessary for understanding mechanisms of response and resistance to cancer therapy. This also is essential for advances in immunotherapy that have recently shown great promise. But these efforts must be done across the different cancer types affecting humankind today.[13]

New methods also imply new regulatory mechanisms for speeding through therapies in shorter time frames. This requires a closer alignment between the science and technology, regulators, and industry. The Food and Drug Administration (FDA) serves as a vital partner in this effort. Once a therapy has been demonstrated to be effective, a willing and able industry partner must be ready to move the technology across the valley of death so it can get into the hands of the patients who need it. Along this path, insurance companies should be incentivized to provide the therapies to patients at reasonable costs.

Government partners have an important role to play in this effort as well. Stable sources of funding will be essential. Promising research cannot be allowed to go dormant due to insufficiency of funds. Public health professionals serving as the first line of defense will need to be adequately resourced and equipped to support prevention, screening, and early detection.

Cancer Breakthroughs 2020 requires skill sets from across multiple communities for the common purpose of eliminating cancer, beginning with prevention efforts, screening and early detection, and available treatments for those who are afflicted. In this way, it reflects both a linear approach to fighting cancer and orthogonal thinking combining a diverse group of experts from across numerous disciplines.

MISSION TO MARS

Launching a successful manned mission to Mars undoubtedly would require orthogonal thinking of the highest order. In some respects, it feels a bit like the Manhattan Project of the twenty-first century. Numerous disciplines would be required to collaborate across the entire spectrum of the scientific and technical fields. Presidents, NASA leaders, scientists and engineers, and business executives have talked about the potential to send humans to Mars.

Previous unmanned missions to Mars have provided an elementary understanding of the fourth planet in our solar system. Space probes have hurled past Mars collecting data, and robotic landing craft have explored small portions of Mars. However, a manned mission would be orders of magnitude more complex. Consider that before 2018, the farthest humans have ventured into space has been to the moon and back. The challenges of keeping a continuous presence in space on the International Space Station (ISS) demonstrate how far humans have come—but also how far we have to go before launching a successful Mars manned mission.

NASA leaders have outlined a series of steps designed to validate their readiness to launch such a mission in the 2030s. The three phases of this developmental process would include:

Earth Reliant (now—mid-2020s): International Space Station operation through 2024; commercial development of low-Earth orbit; development of deep space systems, life support, and human health

Proving Ground (2018–2030): Regular crewed missions and spacewalks in cislunar space; verify deep space habitation and conduct a yearlong mission to validate readiness for Mars; demonstrate integrated human and robotic operations by redirecting and sampling an asteroid boulder

Earth Independent (now—2030s and beyond): Science missions pave the way to Mars; demonstrate entry, descent, and landing and in-situ resource use; conduct robotic roundtrip demonstration with sample return in the late 2020s; send humans to orbit Mars in the early 2030s.[14]

A manned mission to Mars would require scientific discovery and technology development for each of the major systems, beginning with development of life-support systems for a several-year mission to Mars. While humans have lived continuously in space on the ISS since November 2000, the ISS is only a short distance away from Earth should emergencies arise. Explorations into deep space such as a mission to Mars require an inherent independence from Earth that humankind has not demonstrated to this point. R&D will be required to better understand the prolonged effects of cosmic radiation on humans and even on our systems and equipment.

A mission to Mars would require new launch vehicles with greater power to propel heavier loads into deep space. The Space Launch System is the vehicle designed for this purpose. It will need to be accompanied by orbiters, new propulsion systems using solar electric power, transfer spacecraft, landers adapted to the thinner Martian atmosphere, and a habitat for humans for the period the crew will initially be on Mars. Many of these systems are in conceptual stages or in low TRL and will need to reach maturity before they can be confidently included in a future preparatory or Mars mission.

The three sequential phases are each composed of multiple systems and technologies that will need to reach maturity before requirements for the later stages can be developed and validated, adding to the complexity of preparing for a future Mars mission.

A vast array of specialists will need to work collaboratively in highly uncertain environments to make a Mars mission a reality. Along the way, there will be competing demands. How will space on the mission be allocated? What are the trade-offs between supplies, building materials for the envisioned habitat, communications equipment, backup systems for redundancy, and comfort?

Each of the specialists will have good reasons why their component should receive highest priority and be given a greater allocation of space on the mission. The goal must be for each of the specialists to think orthogonally, looking for novel solutions that may be suboptimal in their individual area but optimal from the overall mission standpoint.

CONCLUSIONS

Thinking orthogonally can provide great benefits for technology development, especially for highly complex systems. Having people with a diversity of talents, experiences, and worldviews allows for more robust thinking. Introducing nonlinear or orthogonally oriented thinking also promotes the identification of different paths and multiple solutions that might not have been evident initially.

People with variable risk tolerances can also be very important to finding orthogonal solutions for addressing difficult problems. Structured approaches that cause people to address common problems can promote orthogonal thinking. These approaches are likely to expose a wider variety of linear and nonlinear approaches to problem-solving.

The large, complex examples presented in this section were selected specifically because of the many science and technology areas they touch. However, this is not to say that orthogonal thinking is not useful for smaller daily problems or for the development of individual technologies. Even such simple initiatives as having specialists from other technology fields receive briefings on programs outside of their normal areas can stimulate novel ideas, foster creativity, and promote innovation for both parties.

CHAPTER 10

TECHNOLOGY'S EFFECT ON SOCIETY

Throughout this book, we have been talking about the effects of technology on the course of human development. Whether referring to ancestors 1.76 million years ago or to the Arab Spring, technology has loomed large in all aspects of human life. Through this examination, we have established that the history of humankind and technology have been inextricably linked. However, in this chapter, we will look at the future of technology, in particular considering technology's effect on future society.

Before beginning this journey, some disclaimers are in order. This chapter will not provide an exhaustive list of technologies that are likely to emerge. Literally volumes have been written on this subject, and technologists dedicate their life's work to looking for when the next technology will emerge or where the inflection point in the S-curve will begin. And, of course, getting this right has important financial implications for markets, companies, and investors. This chapter will also not endeavor to rehash dual-use concerns as we look to the future. Suffice to say there are numerous examples of how technologies have been used and misused throughout history. Furthermore, I realize that without careful attention, this discussion could slide off into the world of science fiction, which is certainly not the intent.

Rather, we will take a serious look at what the next inflection point in human development (and by extension technology's development) might look like, how it might come to pass, and perhaps even how to identify the warning signs so as to avoid its pitfalls or be better prepared to respond to them. Such an examination seems prudent and highly useful as we look to the future of technology.

WHAT OTHERS HAVE SAID

Whether technology has been a cause, facilitator, or accelerator of change in society has been part of an ongoing debate. However, not open to debate is the profound effect technology has had on humankind. Technology has allowed humans to achieve a wide range of everyday outcomes that have contributed to the betterment of humankind. Equally clear is that at times, technology has been blamed for outcomes. And at these times, technology has been a source of fear and consternation for many, both for the dangers its misuse could cause and the changes to society that are possible. Some have sought to warn of the negative effects of technology in general and specific technologies in particular. Such concerns have at times come to fruition and at other times been overly alarmist. Despite these mixed outcomes, we are likely to continue to receive such warnings as the march of technology continues.

The Luddite movement in the United Kingdom certainly represented a backlash against the Industrial Age. The Luddites were textile workers who were losing their jobs to the introduction of machinery. They began protesting the new technology in Nottinghamshire on March 11, 1811. The protests eventually spread throughout the northern United Kingdom.[1] While a new Luddite movement has not taken hold today, many have expressed concerns about the pace of technology absorption and the global shifts occurring as a result of technological progress.

The United States has not been immune from such concerns. The thirty-five-year period following the Civil War saw shifting trends in living and working in the United States. One author called it the "death of a rural and agricultural America dominated by farmers and the birth of an urban and industrial America dominated by bankers, industrialists, and city dwellers."[2] Certainly, technology played a significant role in this shift, as yields increased and farming became more efficient, requiring fewer workers. Tariff and trade policies, monetary policies, tax and bank policies, and changes in the structure of farming also contributed. In reaction, farmers developed self-help groups and developed the Farmers Alliance, which became a militant, politically oriented organization for addressing the plight of the farmer.[3]

Just consider technology's effect on global travel and by extension, global travel's effect on humanity over the last 150 years. In the 1850s, circumnavigating the globe was a yearlong trip by ship with great peril. Today, through air travel, this trip can be made in less than twenty-four hours. At the same time, the world's population has grown from three hundred million to almost seven billion people, and these seven billion people are crossing the globe with increasing frequency. Trade, ideas, and even cultures have spread across the globe in these compressed timeframes as well, and at times included elements of the dark side of globalization including societal tensions, open hostilities, and the spread of disease, to name a few.

One analysis highlighting the pace of technological adoption provides a useful point of departure. In comparing the various technologies, the analysis indicated:

- In 1900, <10% of families owned a stove, or had access to electricity or phones
- In 1915, <10% of families owned a car
- In 1930, <10% of families owned a refrigerator or clothes washer
- In 1945, <10% of families owned a clothes dryer or air-conditioning
- In 1960, <10% of families owned a dishwasher or color TV
- In 1975, <10% of families owned a microwave
- In 1990, <10% of families had a cell phone or access to the internet

By 2012, more than 90 percent of the country had these technologies as part of their daily lives.[4] It is important to note both the rate of change of the technologies as well as their transformational nature on society. In this acceleration of technology adoption, these technologies serve a practical primary purpose, in addition to contributing to the development of other technologies.

In such a fast-paced world, it is possible to become consumed by the possibilities. A global society is emerging where ideas move continuously around the world at the speed of light. In this society, we know more about the world and each other than at any time in human history. Technology is also allowing new global communities to evolve that transcend national boundaries. We have seen the tensions resulting from these new relationships, and some have questioned whether traditional governance structures will survive.

The global economy is composed of supply chains that connect raw materials with industrial production, buyers and sellers, and producers and consumers. All are linked in real time by a human-created cyber-domain that fuels a 24-7 economy. The effect on entire industries could be equally profound. By 2018, approximately 10 percent of retail commerce was electronic, which could grow to be as high as 20 percent in the next five years.[5] How will these changes continue to shape the global economy?

The demand for data, information, knowledge, and wisdom continues to grow as well. Corporations, governments, investigators, and police want to capitalize on this digital treasure trove, while regulation on such activity is virtually nonexistent. All of this leaves the individual to decide whether to continue to benefit from this increased sharing despite the inherent downsides.

And while the rapid pace is likely to continue, the shape of the technological advancement is far less certain. In describing the difficulties of predicting the future technology landscape, one futurist lamented, "In the fifties it seemed a pretty much foregone conclusion that by 2015 we would all be commuting to the moon in our flying cars."[6]

While predicting technology's exact path remains a challenge, forecasting effects on society and individuals resulting from future technology seems reasonably possible. On this topic, this same futurist goes on to identify the "10 biggest dangers posed by future technology":

- the total loss of privacy
- a permanent digital connection to the workplace

- all data becoming digital
- having a totally machine based workforce
- the death of human interaction
- the over-reliance on technology
- the increase in technology-related illnesses
- the Gray Goo Scenario [i.e., nanotechnology self-replicating machines take over the Earth]
- the Singularity [i.e., the hypothesized moment when technology radically changes civilization]
- artificial intelligence[7]

Many of the dangers identified correspond to the overuse of information technologies and social media. One warning claimed is that the current information revolution is leading to information overload, that social media can make you feel lonely, and that digital detox is needed.[8] Peter Townsend, who authored *The Dark Side of Technology*, warns that the overuse of Information Age technologies has a negative effect on humans and the infrastructure required to sustain these nearly constant, intrusive communications.[9]

In an interesting turn of events, former Google and Facebook employees have banded together to warn of the negative effects of social media and smartphones. In one case, Google employees circulated petitions to protest accepting artificial intelligence work for the Pentagon that could be used in the conduct of war. One individual has created a new group called the Center for Humane Technology with a focus on the potential for addiction to the technology, particularly in children. Another ex-employee lamented that Facebook is "ripping society apart." As proof of these negative outcomes, these individuals cite the number of times we access our devices, claiming that 40 percent of users access their smartphones ten thousand times per year.[10]

In thinking about the ability of life on Earth to adapt, Charles Darwin's theory of evolutionary biology provides an important starting point. Darwin explained this adaptation beginning from the premise that all life descends from a common ancestor. Darwin theorized that complex creatures evolve from simpler ancestors. Random mutations in genetic code provide beneficial changes or adaptations in what he called natural selection. Darwin's theory postulates a slow but steady natural selection over successive generations.[11]

Brian Arthur's theory that technology was a combination of and descended from other technologies seems to have important parallels with Darwin's theory of evolution, with a significant difference. Scientific discovery, disruptive technologies, novel use cases, and innovation can result from—and, in some cases, seek—a discontinuous outcome rather than a slow, steady process of adaptation.

To illustrate this difference, consider what one author said about the impact of computers on society: "Modern man emerged and began using language 1,400 generations ago. Writing was invented 200 generations ago. Books were first printed 20 generations ago. The invention of the computer occurred less than 2 generations ago."[12]

To put this into perspective, consider that in 1965, Gordon Moore from the Intel Corporation predicted an exponential increase in computing power at fractions of the cost. Moore's law, as it has become known, projected that the number of components on a computer chip would double every two years.[13]

Essentially, during the fifty-plus years since Moore made his prediction, computing power has increased at an exponential rate. The pace of change in computers (and associated IT applications) and the short evolutionary time between generations of information technology platforms means that we may have gone through fifty to one hundred generations of computer technology.

During the same fifty-plus-year period, humans have lived through two generations, arguably increased their mental capacity marginally, improved understanding of the sciences and technology (in large measure due to increases in computing power), and integrated computer technology into society an alarming pace.

In some respects, it feels like computers are to humans what genetically modified organisms are to plants and animals. Just as the use of GMOs in plants and animals allows for introducing traits much more rapidly in successive generations, the use of computer technology, with its exponential rate of change, is potentially affecting humans in this same way. In the case of GMOs in plants and animals, desirable traits have been introduced, at least in theory. However, if the warnings discussed previously regarding computers and social media have any validity, perhaps we have already had undesirable traits introduced into human society through the infusion of internet technologies.

As you are contemplating this possibility, consider: are you a resident of or a tourist in the digital age? The answer to how you feel about the rapid introduction of digital technologies may have a great deal to do with your perspective.

CONVERGENCE OF THREE TECHNOLOGIES AND THE POTENTIAL EFFECT ON SOCIETY

We began this journey with the promise that a list of technologies identifying those technologies that are likely to be the most important over the next five, ten, twenty-five years would not be provided. We will hold to that promise, but it is important in thinking about how technology could affect humankind and society to examine a few technologies that, when combined, could affect the essence of what it means to be human. The three technologies are biotechnology, artificial intelligence, and the Internet of Things. Other technologies such as quantum computing, mobile internet and cloud computing, advanced energy, advanced materials, robotics and autonomous systems, and 3-D printing will undoubtedly be critically important as well. In fact, one cannot begin to understand why biotech, AI, and the IoT could have such a pivotal role for humankind without also talking about these other technologies. However, these three, taken together, could play a unique role in altering the very essence of human existence.

Throughout this book, we have been talking about the idea of the convergence of technologies. We have repeatedly asserted that individual technologies are less important than how technologies are combined with other technologies to create larger technologies and even systems. Before discussing the convergence of these three technologies, we will look at them individually to gain a better appreciation for the potentially life-altering changes the convergence of these three technologies could have on individuals, societies, and the species.

BIOTECHNOLOGY

Biotechnology is undergoing a significant transformation that is unmatched across any other technology area. While computers and IT have been governed by Moore's law, a similar analysis was done concerning biotechnology. The result is what has become known as Carlson's curve.[14] Dr. Rob Carlson demonstrated that increases in productivity for DNA sequencing and synthesis, combined with cost decreases, have made this biotechnology more widely available; and in fact, the changes in biotechnology that Carlson cataloged were even more significant than those in the IT sector.

While Carlson focused on DNA sequencing and synthesis, the 1998 Congressional Office of Technology Assessment analysis that was cited previously assessed the rate of change of fourteen biotechnologies over the sixty-year period from 1940–2000. The finding was that rates were increasing exponentially and doubling in capacity every six months, or 400 percent per year.

Since the discovery of the three-dimensional structure of deoxyribonucleic acid or DNA in 1953, there has been a quest to understand its properties, its relevance to defining life-forms and their characteristics, and the potential to alter DNA sequences.[15] Through this work, an entirely new field of study arose centered on the manipulation of DNA, first with recombinant DNA engineering and now progressing to synthetic biology.

The positive and negative implications of DNA engineering were immediately recognized. Experts quickly realized that the same technology used to eliminate undesirable traits in a genome could also be used in dangerous ways. The Asilomar Conference on Recombinant DNA in 1975 brought together more than 140 biologists, lawyers, and physicians to discuss the potential biohazards and the regulation of this emerging biotechnology.

Over the subsequent forty-five-year period, major advances have occurred in the development of key technologies. Polymerase chain reaction (PCR) was developed in 1985 for amplifying DNA sequences. The Human Genome Project (1992–2003) decoded the entire human genome. Building on these technologies, a synthetic polio virus was developed (2002), a program called the Encyclopedia of DNA Elements (ENCODE) (beginning in 2003) was developed for cataloging and understanding the human genome, the first synthetic cell was developed (2010), and Clustered Regularly Interspaced Short Palindromic Repeats (CRISPR and CRISPR-associated [Cas] genes [CRISPR-Cas9]) was developed as a tool for a range of synthetic biology applications. And it is the CRISPR technology that has made gene editing "faster, cheaper, more accurate, and more efficient," as well as more accessible. The economic benefits of gene editing technologies are clear as well, with one prediction that CRISPR alone will be a $10 billion market by 2025.[16]

What is interesting about the synergy between advances in biotechnology and information technology is that they are occurring in unison. Taken together, these applications provide the necessary tools, information, and capabilities to understand DNA in ways that heretofore have not been possible. And with these new understandings, it becomes possible to do informed and targeted manipulations of an organism's genome.

For example, PCR allowed for amplification of DNA and RNA, diagnosis of genetic defects, bioforensics, and evolutionary biology studies.[17] Without the science and knowledge generated by the discovery of PCR, the development of a synthetic virus by Dr. Eckard Wimmer from the State University of New York could not have occurred.[18] The first synthetic cell—a goat parasite called *Mycoplasma mycoides*—relied on knowledge gained though the development of the synthetic polio virus. It also relied on advances in genomic sequencing, which made the process faster, more efficient, and more precise.[19]

The Human Genome Project and ENCODE specifically employed IT and big data applications—now called bioinformatics—to improve understanding of the human genome. The Human Genome Project sequenced the 3.3 billion base pairs of the human genome. ENCODE was designed to interpret the sequences, providing knowledge about the function of the other 99 percent of the human genome whose purpose remains unknown.[20] CRISPR-Cas9 provides tools for DNA manipulation and gene editing that have taken the ease of use and precision of the process to an entirely new level. Undoubtedly, more advances are in the pipeline.

The potential benefit for these technologies is evident. Human quality of life and longevity are likely to increase. Personalized medicine and genetic engineering could deliver targeted therapies. Acute and chronic diseases that once plagued humankind will likely be eliminated. Biotechnology tools will likely increase agriculture capabilities, leading to higher yields, higher quality, and more nutrition in our food—all at lower production costs. Biofuels have the potential to provide alternative means for fulfilling human energy needs.

But biotech has a potentially dark side. The increasing availability of technologies such as CRISPR will provide increasingly sophisticated tools in the hands of those who may lack the fundamental understanding of the science, the potential for harm, or the dangers of misuse. The outcomes could range from accidents and negligence to the deliberate use of biotechnology in developing dangerous pathogens or even species-altering changes to a life-form.

Equally likely are concerns that biotechnology experiments for legitimate purposes—some for applications that are not fully understood this time—are more regularly being conducted. In 2017, a report surfaced that a three-person embryo had been created. The embryo used DNA from a healthy donor, placed in an egg cell from a woman with a neurological disease. This substitution left the donor's healthy mitochondria intact. The modified egg was then fertilized and implanted into the uterus of the mother. While there were no complications in the procedure, worries remain about the long-term health of the child given the mismatched DNA and mitochondrial DNA, which has been demonstrated to be detrimental in the past in animal studies. In one experiment using mice, mismatched DNA led to adults with reduced physical performance, learning, and growth.[21] Concerns about this particular procedure were increased because some of the DNA carrying the mother's disease was inadvertently carried into the donor egg.[22]

Genetic modification, or enhancement, has also received attention recently. Substituting mitochondrial DNA or using gene editing to cure diseases such as hemophilia and sickle cell anemia could have great societal benefit, with the potential to confer important benefits for the individual by returning her to a "healthy condition." However, genetic modifications also have the potential to be used to confer preferential traits such as hair and eye color, enhance levels of performance, increase muscle mass, and delay the aging process.[23] In the case of the latter enhancements, these modifications would be based on preference rather than medical necessity.

Going back to our discussion of Darwin and evolutionary biology, the point was made that modern humans emerged and began using language over the course of 1,400 generations; however, through gene editing, one could foresee a discontinuous evolution of the human genome that could occur between successive generations using gene editing techniques that would make fundamental changes to the germline.

And it is not just the human genome that could be threatened. In 2017, sixteen European nations voted against the use of GMO crops for planting or consumption. Some of the nations expressed concerns about GMO crops and animals for human consumption and as part of the environment. One report indicated that genetically modified crops could unintentionally kill butterflies and moths.[24] To date, little tangible evidence of harm has been identified, and the increased yields and efficiencies seem to outweigh the negatives. Still, many are concerned about the long-term implications of the use of GMO in agriculture.[25]

Before moving on, it is useful to consider some of the important questions that surround the use of genetic modifications on living organisms. What are the implications of such changes? Is the science well enough understood to allow for such manipulations to take place? Do we understand the range of potential outcomes of such changes? Are there some experiments and modifications to a genome that should definitely not be undertaken under any circumstances? What are the potential implications of such a dramatic and instantaneous shift in the human genome?

ARTIFICIAL INTELLIGENCE

Artificial intelligence continues to be a source of promise and concern for society. Some see the benefits of AI, allowing for reaching new heights as these capabilities become incorporated into all aspects of society. Others see only the dangers of a workerless society controlled by computers. And still others take a laissez-faire approach to the emerging AI era. Regardless of where one lies on the spectrum, many understand that AI and its associated technologies are likely to result in an inflection point in the history of humankind.

As in many fields of study, there is no agreed definition of AI across the community. In fact, there is not even a single community when it comes to AI. Often AI is used in conjunction with or interchangeably for robotics and autonomous systems. However, this would be a great misrepresentation of AI and its potential for affecting society and individuals. Given this introduction to the topic, autonomous systems (and robotics) and AI should be defined for our purposes going forward.

We have used a shorthand by placing autonomous systems and robotics in a single category. Robots have been around for some time, dating back to the concept of a mechanical man, the anthrobot, designed by Leonardo da Vinci. Robots have been used regularly since 1961, when the first industrial robot was incorporated by General Motors into an automobile factory in New Jersey; the function of this robot was to do spot welding and extract die castings. In comparing robots and humans, one account states, "Robots are better at crunching numbers, lifting heavy objects, and in certain contexts, moving with precision.

Humans are better than robots at abstraction, generalization, and creative thinking."[26]

Autonomous systems, on the other hand, act independently of controllers and can respond as programmed to stimuli. In this sense, an autonomous system can really be thought of as a robot with the ability to act independently and respond to outside stimuli. In other words through a system of onboard sensors, the autonomous system makes "decisions" based on external stimuli.

Turning to AI, the *English Oxford Living Dictionary* defines it as "The theory and development of computer systems able to perform tasks normally requiring human intelligence, such as visual perception, speech recognition, decision-making, and translation between languages." Amazon defines AI as "the field of computer science dedicated to solving cognitive problems commonly associated with human intelligence, such as learning, problem solving, and pattern recognition."[27]

The two definitions have similarities in computer systems and intelligence. Still, it is not clear that either of these two definitions quite explains the novelty of AI. A more straightforward way to think about autonomous systems versus AI is to think of autonomous systems as "making decisions" based on programming that relates stimuli with actions taken, while AI begins with a set of stimuli-response sets but has the capacity to "learn."

A shorthand way to think about the difference between autonomous systems and AI is that autonomous systems depend on regression analysis and therefore only act within a defined set of parameters, while AI allows the system to act within both programmed and learned parameters.

The taxonomy of AI includes symbolic learning, based on image-processing technology and machine learning, that is heavily reliant on data and algorithms, essentially to teach the computer the patterns necessary for making predictions. Machine learning includes statistical learning and deep learning. Statistical learning techniques are used in speech recognition and natural language processing, relying on massive amounts of data and algorithms for drawing conclusions. And it is the second branch of machine learning called deep learning that both excites and concerns many.

Deep learning entails the development of neural networks— inspired by the biological neural networks in the human brain—that many experts believe can mimic the brain's capacity for learning. Deep learning is based on learning data representations as opposed to employing task-specific algorithms. The implication is that instead of being programmed every step of the way, the computer has general instructions that allow it to learn from data without new step-by-step instructions. Should this promise be realized, AI could provide human functionality in silicon, in a synthetic environment.

So using the deep-learning functionality of neural networks, AI relies on programmed stimuli-response patterns but continues to learn and store these new experiences for use in the future when the stimuli are confronted again. This learning is done using specially designed, complex neural networks that in theory replicate how the human brain functions. In mimicking the functionality of the human brain, AI also promises to allow humanlike reasoning in nonhuman computers.

Some have debated whether such humanlike qualities are possible from a computer. Others see it on the near horizon. In 2017, Google's DeepMind program "decimated" AlphaGo world champion Ke Jie using learning from expert human moves and reinforcement learning through self-play.[28] Next, DeepMind developed a new version of the game (AlphaGo Zero), which it mastered rapidly; in fact, when pitted against an older version of itself, it won one hundred games to zero."[29]

DeepMind's program also produced interesting unexplained findings. The program appeared to be developing and implementing strategies that human players had never seen before, leaving some to wonder if actual learning had taken place. Strategies were at times defined as "madcap to human players," unorthodox or even incomprehensible.[30]

Many possible scenarios have been discussed regarding the future of AI. The ranges go from benign to malevolent. In the benign world of AI, the technology is used to assist humans and perform tasks that are dangerous or that humans do not want to do. In the malevolent world of AI, computers take over the world. They make other autonomous systems, develop independent strategies, and take over the planet to the detriment of humanity. Neither of these cases appears to be very likely outcomes for the future of AI.

Let's look at what we know today. Automation and robotics have been slowly encroaching into various aspects of our lives. Many highly repetitive factory and industrial jobs previously done by humans have been taken over by robots and autonomous systems. This includes jobs in manufacturing, mining, and even services; we are even seeing positions at restaurants being dehumanized, for example the placing of an order. We should expect these trends to continue in areas where it makes economic sense to do so, where sufficient skilled labor is not available, or where there are safety concerns.

While this substitution is occurring, another complementary trend is the rise in productivity. General Motors lost one-third of its 600,000 workers since 1970 yet is outproducing its previous production totals. The steel and metals industry has lost more than 265,000 jobs, or 42 percent of its workforce, yet production has increased 38 percent. One estimate is that by 2020, the United States will be the "most competitive country in manufacturing."[31] While labor in mining has been reduced from more than 250,000 to 58,000, new methods of mining that are less human-intensive have been discovered.

The maturing of autonomous vehicle technology will likely usher in a massive restructuring in the economy. Long-haul trucking, which is a dangerous and tedious job, is likely to be one of the first to go. One can easily envision autonomous trucks traveling down the highway in convoys across the country. Given that some 15 million US workers have some relationship with the industry (including approximately 3.5 million professional drivers and 5.2 million more in the industry in nondriving positions, and another 7 million or so spread throughout the federal highway system who support the trucking industry),[32] many of these positions are likely to fall to autonomous systems. Of note— given that the US workforce is approximately 153 million people, this

15 million in the trucking industry represents 10 percent of the total workforce.[33]

Over time, additional uses of AI will likely be identified, and the displacement will continue. As the technology for autonomous vehicles matures, we should expect to see a restructuring of the economy in several areas. Driverless cars look to be on the near horizon, which could alter many industries and even change the way we think about car ownership. One account optimistically predicts autonomous vehicles will "prevent accidents, ease traffic and bring newfound mobility to the elderly and disabled."[34] It could also save 1.2 million lives annually—forty thousand in the United States alone—from losses due to vehicle accidents.[35] Of course, we cannot get to this place until the legal, ethical, and policy issues have been ironed out; the technology has matured; and humans become comfortable with a driverless world.

We should expect that the effect of autonomous systems on the future of work will be profound. As autonomous technologies mature and gain greater performance that matches or exceeds human capabilities, a higher percentage of the workforce will either be displaced or at risk. One estimate suggests that automation technologies could affect 50 percent of the world economy, or 1.2 billion employees and $14.6 trillion in wages.[36] When these people are displaced, what will they do to earn a living and provide for their families?

Given the nature of technology, we should expect that autonomous systems will begin to proliferate—perhaps more slowly and less dramatically than some have offered, but these technologies will still also likely be far more disruptive than some of the naysayers have made it seem. To be sure, the adoption of autonomous systems will not be without concerns and controversies.

For example, going back to the question of autonomous vehicles, major concerns include the safety and decision-making of the systems. Despite the loss of almost forty thousand Americans a year in car accidents, largely due to their own mistakes or negligence, many are concerned that offloading the responsibility for decision-making to machines places humans at risk through no fault of their own. Algorithms would calculate preferential loss such that in no-fault accidents involving two cars where the occupants of one of the cars are likely to die, the decisions would be based on the number of occupants in each vehicle. Therefore, the vehicle with three people would have preference in the algorithm over the vehicle with only a single occupant.

Regarding autonomous vehicles, we are entering a dangerous period in which the fleet of vehicles on the road is likely to become a mix of human-controlled and autonomous vehicles. In such a case, the danger increases, as the highly predictive autonomous vehicle algorithms and the highly individual, probabilistic human driving reactions and habits are likely to collide, both figuratively and literally. Furthermore, when society begins widespread deployment of autonomous vehicle technology, an inherent assumption will be that the system has great accuracy and surety in all communications between vehicles and control systems.

Contemplating the use of autonomous systems in warfare also leads to some interesting questions. For decades, the military has employed robotics to assist with certain routine tasks; an example is the use of robotics for loading artillery projectiles. However, now broader uses of robotics on the battlefield are being considered to replace humans in many critical areas.

DARPA ran a grand challenge for autonomous vehicles on the battlefield in the late 1990s, requiring teams to navigate a closed track that included obstacles and a requirement to travel in a convoy of vehicles. Robots for autonomous systems have been embedded in such systems as naval guns on ships and even artillery for more than a decade. Robotic carriers for transporting individuals and supplies have been contemplated or are under development as well.

With advances in AI, it is now possible to envision a future where the entire engagement process—what the military calls the "kill chain"—would be embedded in a single platform with no human involvement. This means that the entire decision-making chain from observing to destroying the target would be done with no humans in the loop.

A Defense Science Board considered the uses and possible concerns associated with the incorporation of autonomous systems into military operations. The study concluded that there could be benefits for the warfighter using autonomous systems for particularly dangerous missions; that research on autonomy needed to continue to better understand what adversaries are or could be using AI for; and that certain decisions to be made on the battlefield, particularly those involving the use of deadly force, require a human brain.[37]

A couple of important notes are in order regarding autonomous systems in battle. From the earliest times, warfare has been regarded as a human endeavor to be undertaken only when the interests of the state are at risk. It has represented a conflict between humans in which the losers—and even many on the winning side—died while fighting for their objectives. To dehumanize warfare has the potential to make wars between nations more imaginable and therefore more likely. In addition, the use of autonomous systems against human formations presents ethical and moral dilemmas that will need to be contemplated.

A range of questions come to mind when thinking of the possibilities of AI and autonomous systems. What are the limits of AI and autonomous systems? Do either of the alternative futures presented earlier reflect reality for AI? Is it possible to establish limits into which AI should not be allowed to encroach? Are there ways to gain the benefits of AI while maintaining what it means to be human? Are there areas of human endeavor where AI and autonomous systems should be restricted, either through national laws or international agreements? Will specialized legal instruments be required to adjudicate AI infractions and issues?

INTERNET OF THINGS (IOT)

One cannot begin a discussion of IoT without first introducing some of the features of the internet. The internet was created in 1969 at the University of California, Los Angeles. Its purpose was to serve as an information-sharing platform. In the early days, meaning through the mid-1980s, there were fewer than ten thousand users, most of them dedicated to sharing information on research.[38]

However, there was an inherent dilemma with the internet. When it was first designed, the goal was to have an open architecture that would facilitate information sharing; in contrast, a proprietary or closed system would have been counter to the goals of the internet. It was also designed with virtually no governance, other than for technical control of the system and naming conventions. There was no deliberate architecture for what has evolved into this new domain we call cyberspace. Furthermore, the early pioneers of the internet did not design security features into the system, as they were perceived to not be necessary.

Today, the internet has grown into a global network that has become embedded in virtually all aspects of human life. It is estimated that at the end of 2017, more than 4.0 billion of the 7.6 billion people on the planet, or almost 52 percent of the world's population, were Internet users.[39] The once small information-sharing platform has grown into something much more, and while information sharing continues to be a central feature, the internet now more resembles a utility company in many regards. Hanging on the end of the network and within the cloud, one sees infrastructure, security and surveillance, transportation, industrial controls, healthcare, retail services, and even home controls.

So what is the IoT? In describing the IoT, one source states,

> A thing, in the Internet of Things, can be a person with a heart monitor implant, a farm animal with a biochip transponder, an automobile that has built-in sensors to alert the driver when tire pressure is low—or any other natural or man-made object that can be assigned an IP address and provided with the ability to transfer data over a network. . . .
>
> IoT has evolved from the convergence of wireless technologies, micro-electromechanical systems (MEMS), microservices and the internet. The convergence has helped tear down the silo walls between operational technology (OT) and information technology (IT), allowing unstructured machine-generated data to be analyzed for insights that will drive improvements.[40]

By 2020, it is estimated the IoT will have more than twenty-five billion multipurpose devices. Early IoT devices in 2000 included radio frequency identification (RFID) tags for routing, inventory, and loss prevention for supply chains. In the mid-2000s, IoT devices began to move into the market space with increasing frequency for surveillance and security cameras, transportation safety, food safety, and healthcare applications. Throughout the next decade, applications have continued to grow for refrigerators, washing machines, cars, smart buildings, and even smart cities. As these devices continue to get smaller, more capable, and more energy efficient, they will undoubtedly find additional uses throughout society, government, and industry.

It is not difficult to envision that within the next decade, one will find it nearly impossible to be out of range of some sort of IoT device. Getting off the grid will not be possible unless one is willing to completely forgo all modern conveniences and human interactions. When this IoT proliferation occurs, significant effects will be felt concerning basic human privacy.

The IoT will also be used to control the daily aspects of our lives. We will effortlessly and frequently go online for all matters of activity. However, we should also consider whether this confidence is somehow misplaced. For just as with the internet, the IoT was not designed with security considerations first but, rather, as a means of connecting electronic systems for information-sharing purposes. To emphasize this point, consider that in 2016, the Department of Homeland Security issued guidance for how to think about IoT security, highlighting six areas for improvement: (1) incorporating security at the design phase, (2) advancing security updates and vulnerability management, (3) building on proven security practices, (4) prioritizing security measures according to potential impact, (5) promoting transparency across the IoT, and (6) doing so carefully and deliberately.[41]

One would have hoped that such guidance had already been thought of in the early days of the IoT, rather than as we approach twenty-five billion IoT devices that have already been fielded.

In considering the implications of the IoT, understanding how it will affect privacy is essential. Today through constant connectivity in the form of smartphones, smart watches, wearables, embedded electronics, and third-party sensors (such as video cameras), our every move can be known and tracked in real time. The interesting thing about this phenomenon is that humans adopted this capability, even though it runs counter to many people's desire for privacy.

As we move farther into the IoT world, sensors and devices will always be present, brokering data around the globe. They will be capturing and sharing data 24-7 to provide the potential for nearly unlimited and instantaneous situational awareness and communications. For those who opt in, they will be tethered to the internet both consciously and unconsciously. For those who do not opt-in, their data will likely be captured anyway as they interface with others who have joined the IoT revolution.

The IoT will also serve as an accelerant. As news and information are beamed around the world, humans will always be just milliseconds away from the next news cycle. Social media will provide the forum for generating physical and virtual protests instantaneously. Such speed will far surpass the speed of lawmakers, policy makers, and law enforcement to react. In effect, an impedance mismatch will occur between authorities, who operate at a nineteenth-century pace of lengthy legislation development and long debates, and the twenty-first-century forces generated by social media, which will be agile and highly responsive.

AT THE INTERSECTION OF BIOTECHNOLOGY, ARTIFICIAL INTELLIGENCE, AND THE INTERNET OF THINGS

In the beginning of this convergence section, the point was made that changes in biotech, AI, and the IoT, either individually or combined, could affect the essence of what it means to be human. That is a powerful assertion and certainly one that should be explained in greater detail.

Three qualities that define what it means to be human are the physical, mental, and emotional states of being. Changes to any one of these states could have significant impacts on humankind. And potentially, changes to one or more could alter what it means to be human.

In the case of biotechnology, the physical and emotional characteristics of humans could be redefined by changes to the human genome
and could dangerously alter human life and the environment. Darwin's
theory of evolutionary biology accounts for a natural selection whereby a
species continues through the accumulation of traits that promote survival. The process is continuous, meaning that the traits of a species are
passed from one's ancestors through an evolutionary process. Employing
biotechnology to edit the human genome creates the potential for a discontinuous germline in which the genomic material has been inserted
rather than evolved and is then passed to successive generations.

The broader effects of such manipulations would likely not be fully
known at the time, given that today we only understand 1 percent of the
function of the 3.3 billion base pairs that make up the human genome.
Obviously, the degree to which a discontinuous germline is created would
depend on the numbers and types of genetic changes that are created.
Still, even "minor" manipulations should receive considerable scrutiny to
ensure the science is understood and the risks are minimized.

AI similarly has the potential to alter the human experience. A
large part of being human is the ability to perceive, think, and reason.
If true artificial intelligence can be created, what does that say about the
human experience? As machines begin to replace humans in all matters
of activity, from heavy lifting to tasks requiring manual dexterity to
higher-order reasoning, humans could become irrelevant and perhaps
even replaceable. From a more practical standpoint, this substitution
of humans for AI and autonomous systems presents problems for what
to do with displaced workers. From the earliest days of humankind,
when survival was a very real issue, activity that we came to call *work*
provided a purpose. The use of robotics and automation has served to
create greater effectiveness and efficiency in the tasks that humans can
perform. In the future, it is entirely possible that a large majority of
workers could be displaced by AI-enabled robots and automation.

Finally, the IoT—enabled by communications, facilitated by the
cloud, and willingly invited into our lives—will redefine privacy for
humanity. Expectations of personal privacy can be seen in the earliest

structures that humankind created. Dwellings included multiple rooms that created privacy for their inhabitants. Democracy is based on the premise of individual liberties, freedom, and privacy in our daily lives. With the potential for all movements to be monitored and tracked, there will be a corresponding loss of privacy. It will likely not be long into the future when such monitoring could use facial microexpressions or perhaps even brainwave activity to assess individuals. While in both cases, sensors to measure these changes and an understanding of how they relate to activities and behaviors would need to be correlated, private thoughts and feelings could cease to be private should this occur.

Concerns about technology reaching a hypothetical point at which it fundamentally alters humanity are not new. John von Newmann— the genius mathematician, physicist, and computer scientist who made many contributions across numerous fields of science and technology— is credited with having first coined the term the *singularity* to describe this theoretical future point. Others have written extensively on the topic as well, cautioning that the singularity would result in a condition that is both uncontrollable and irreversible. At the core of singularity is the concern about the effects of creating AI-enabled computers that have superhuman intelligence.

Imagining a world in which the meaning of humanity has been altered should provide cause for concern and reflection. Never before has humankind had such capacity. How we navigate through these challenges and the accompanying moral and ethical dilemmas these advancements could pose has implications for the society that will remain and the future of the human experience.

This section on convergence has two primary purposes. The first was to identify technologies that have the potential to fundamentally alter human existence. The second was to demonstrate the effects of convergence of technologies—how the whole is far greater than the sum of the parts. In no way was this meant to be alarmist. Rather, it is hoped that through raising these issues and the potential pitfalls, considerations can be given to how to "manage" these technologies and avoid their potentially adverse effects on society and the individual.

TOWARD A TECHNOLOGY-SOPHISTICATED WORKFORCE

Throughout this book, much attention has been paid to the development of technology and how technologies are changing society and individuals. Yet technology proliferation is also having a profound effect on the workforce. In this section, we want to consider those changes and think about what might be the skill sets required for a worker in this technologically enabled world.

Remembering the four generations of industry R&D discussed previously can serve as a useful point of departure. First-generation companies conducted their R&D internally. The second generation allowed for the increasing complexity that required specialists for certain technologies, the third generation saw management making conscious decisions to focus on core competencies with outsourcing for specialties or cost savings, and the fourth generation was based on the decision to use strategic technology sourcing for understanding of and access to technologies that could be brought forward for use. Given such a paradigm, what does this say about the type of workers that would be needed for technology development in the future?

Obviously, scientists will still be required for basic and applied research and early development. Likewise, operators will still be essential for identifying practical problems to be solved. Linking the scientists and operators will be technologists with a foot in both camps, an understanding of the science and the operational problem as well as the ability to bridge the divide between the two to communicate between them.

Given the increasing specialization of technologies, where technologies are combinations of other technologies, more of the workforce will need to be dedicated to interfacing with people in highly specialized fields and doing research that may not be visible if one only looks at business intelligence research and databases.

Such interfacing will require people with a broad mix of skills—some might even say Renaissance men and women. They will need to be equally conversant in areas such as metallurgy and 3-D printing, cell structure and biotechnology, packet switching and the internet, and policy and ethics. Having such diverse skills will allow them to understand the convergences of technologies that will drive change and ensure that the whole is far greater than the mere sum of the parts.

These new technologists would do well to have a broad background, including in science, technology, engineering, and mathematics (STEM) fields, and be equally adept in social science areas. They will need to be connectors who are always on the move, looking for new types of and ways to use technology. For personality, they will need to be innovators who have vision and an entrepreneurial spirit.

Developing teams will require much the same inclusiveness as on the Manhattan Project decades ago. Groups of individuals with specialized skills in the sciences, technology, and operations will need to be brought together. Crossfunctional teams composed of people from varying backgrounds will be essential to solving the technology problems of tomorrow. These teams will be a mix of the virtual and physical and will likely include combinations of AI or autonomous systems and humans, taking advantage of the strengths of each while mitigating the weaknesses.

The common thread uniting the crossfunctional teams will be a sense of mission and a common purpose. Despite diverse backgrounds, leaders must be able to harness the talents of all for this common purpose. At the same time, in this technologically enabled world, our concepts of leadership may need to change as well. Rather than thinking of a central leader or hierarchy, it will be necessary to think of networked leadership where each of the elements operates independently based on a common vision or goal.

CONCLUSIONS

If history is any indication, technology will continue to play a symbiotic role with the development of humankind.

Since early development, humankind has built cities and societies with technology as a constant companion. Whether humankind's quest for advancement or technology's demonstrated possibilities led or followed in this developmental process is sometimes difficult to ascertain. The only thing that has been certain is the continuous advancement of technology.

Today, through globalization, the world has been made smaller, we know more about each other than at any point in history, and the march of technology continues at a breakneck pace. Whether technology will continue to bring societies closer together and advancements will continue along an exponential development path remains an open question. One could easily postulate alternative futures where virtual communities replace physical ones, leading to the compartmentalization of societies or perhaps even a Luddite-inspired backlash that sees retrenchment in technology progress. While such discussions are beyond the scope of this book, they provide interesting grist for thought.

CHAPTER 11

THE TECHNOLOGIST'S RESPONSIBILITY TO SOCIETY

This chapter is designed to speak directly to the technologist—to those leading and pursuing the scientific discovery and technological advances—to provide food for thought concerning the future development of technology. Undoubtedly, as technology begets more technology, there will be opportunities to progress even further across a broad range of technology areas.

However, before pursuing such advances, perhaps the scientist, technology developer, or engineer will be tempted to pause to reflect upon his responsibility to society. Just because something can be done, does it mean it should be done? Such a reflective question is not meant to retard science and technology, research and development, or innovation or transformation, but rather to ensure that discovery and development proceed in an appropriate manner that considers the established principles and best practices that have come to be expected of the scientific and technical communities.

Tools for managing technology including the moral, legal, and ethical issues associated with technology development have been discussed previously. This chapter will consider these issues from the perspective of the technologist. Previously, our discussions highlighted institutional processes that were imposed. In contrast, this chapter will consider technology management from the standpoint of what is required of the individual.

Invariably, a new technology comes with instructions accompanied by caveats and disclaimers. These documents are intended to establish the legal mechanisms and framework to protect the developers and producers of the technology. They imply situations that might occur and risks that might result from the use of the technology but make little attempt to quantify the likelihood of such negative outcomes. They might also include the expected operating conditions for which the technology has been designed; the implication is that if the technology is being used outside of these parameters, then they make no claims about its ability to operate safely or as intended.

While such a narrowing of the operating parameters might be appropriate for certain technologies, it likely fails to account for events—either naturally occurring or manmade—that might occur at the tails of the distribution curves. For example, infrastructure projects are not normally constructed to withstand the results of a hundred-thousand-year flood. Doing so would be prohibitively expensive and still not ensure the viability of the technology (in this case, the critical infrastructure) in an extreme weather event.

Given this backdrop, the technologist faces a very real conundrum. Does she design for the technology to operate on the extremes or focus on the most likely scenarios? What degree of surety should be built into technologies? And what happens if the assumptions and planning factors are wrong and fail to account for key scenarios and thus potentially likely operational outcomes?

DESIGNING TECHNOLOGY IN UNCHARTED TERRITORY

The history of technology development is about pushing beyond design limits. That is why technology development is a nonlinear process with many stops and starts, much trial and error, and numerous experiments before a technology is deemed operationally ready to be employed. The degree to which a technology is subjected to challenging stress tests increases the likelihood it will not fail (assuming, of course, that it has been modified following testing to fix design flaws and is being operated as intended).

Let's begin this discussion by considering the technologist's role in developing a technology. Based on a practical need, a technologist—perhaps based on direction from a boss or on an observed capability shortfall—develops a technology. The technology performs as intended and works within the operational parameters that have been established. Appropriate safeguards are established to protect against the misuse of that technology. Perhaps the technologist applies for and receives a patent to protect the IP. All seems to be going well; the product or capability sells and begins to proliferate.

But then something occurs that the technologist failed to anticipate. As the technology gains more widespread use, other uses are discovered, or perhaps individuals not intended to use the technology become widespread adopters. The Kalashnikov rifle provides an example of such a technology.[1]

The rifle was developed by Mikhail Kalashnikov while he lay in a hospital recovering from wounds sustained in combat during World War II. The first Kalashnikov—or AK-47, as it is commonly referred to—was fielded in 1947 and has become the staple for the Soviet and now Russian armed forces, their allies, paramilitary forces, and private citizens. Militaries from 106 countries around the world use the AK-47 due to its simplicity, low production costs, and reliability. One ominous account states, "AK-47s have caused more deaths than artillery fire, airstrikes and rocket attacks combined."[2]

With such a record of proliferation and destruction, what degree of responsibility should be assumed by its inventor? Should Mikhail Kalashnikov have been held responsible for the results of the technology that bears his name? Or does the technologist's responsibility end with the successful development of a technology? In this case, it seems unreasonable to hold the inventor responsible for all use cases and the subsequent killings that have occurred throughout the globe.

What about the case of the *Titanic*, which was designed as and claimed to be unsinkable? Should the naval architects have envisioned its operation in extreme conditions (including low temperatures and high-impact loading) that affected the structural integrity of the hull? The fact remains that numerous examples can be found where technologies are operated beyond the established and verified design parameters. Many occur without any adverse impacts. In some cases, these are documented and lead to new use cases and even greater proliferation of the technology. In others, the technology fails, sometimes with catastrophic consequences. It is instructive to examine some of these cases to see what can be learned from them. Also instructive is to observe the reaction of the technologist once a "flaw" has been discovered.

The *Deepwater Horizon* oil spill, which began on April 20, 2010, provides one such example. The *Deepwater Horizon* was a mobile, floating, dynamically positioned drilling rig that could operate in ten thousand feet of water.[3] It was operating at the very edges of or beyond its operational limits, relying on concrete that had been cured using new techniques, when high-pressure methane gas from the well rose into the drilling rig and exploded. The spill lasted for eighty-seven days, releasing up to five thousand barrels per day for a total of more than 4.9 million barrels of oil. The disaster was the single worst US oil spill.

The case was taken to federal court, where a judge ruled that the blame was to be shared among three companies. British Petroleum (BP) was found to bear 67 percent; Swiss-based drilling rig owner Transocean accounted for 30 percent; and Houston-based cement contractor Halliburton Energy Service took 3 percent of the blame. The judge also found that BP's "profit-driven decisions" during the drilling of the well led to the deadly blowout. The companies face civil litigation as well, which could cost BP almost $20 billion by one estimate.[4]

While a series of errors and operational decisions eventually led to the failure, several technologies failed simultaneously, contributing to the catastrophic outcome. The concrete blowout protector failed, as the concrete had not cured. The rig was being operated at depths and over distances that were at the edge of the operational envelope. The plans for control and remediation after a blowout were not adequate to contain the oil spewing from the out-of-control well.

One account lists eight failures of "safety systems" intended to prevent this kind of incident: dodgy cement, a valve failure, pressure test misinterpreted, leak not spotted soon enough, a second valve failure, an overwhelmed separator, no gas alarm, and no battery for the blowout preventer.[5] Given these individual and ultimately cascading failures, the *Deepwater Horizon* oil spill outcome should not have come as a surprise.

The incident also established the limits of what government can reasonably be expected to do in such circumstances. While the government can enforce laws and make regulations, it does not have the technical capabilities to respond to an incident of this magnitude. The government had to depend on industry to stop the uncontrolled flow of oil into the Gulf of Mexico.

FAILURE TO DESIGN TECHNOLOGY WITH SAFETY AS A MANDATORY FEATURE

Thinking back to the earlier discussions of the scientific method and technology development, the methods were designed to develop repeatable outcomes confirming or refuting the stated hypothesis. Thus, far less attention is normally applied to testing the possibilities or risks of adverse outcomes resulting from operating outside of the operational envelope for which the technology has been designed.

A notable exception has been in the development of medical treatments, including vaccines, drugs, and procedures, where efficacy and safety must be demonstrated to provide a large therapeutic window.[6] In the development of vaccines, and other medical treatments ingested by humans, efficacy and safety carry roughly equivalent weight.

However, in other technology fields this is not always the case. And the result of considering safety almost as an afterthought is that far less energy is generally spent thinking about the potential risks, dangers, and misuse of the technology than the functionality and use cases for which the technology was originally developed.

Several nuclear accidents highlight failures to design technology with safety as a mandatory feature. In each of these cases, other design flaws and human error also played a key role in the catastrophic failures that ensued.

On March 29, 1979, the Three Mile Island Unit 2 reactor experienced a partial meltdown and subsequent radiological release. The accident was attributed to a combination of "equipment malfunctions, design-related problems and worker errors."[7] While only a small amount of radiological material was released and the resultant health effects were negligible, the failure generated considerable concern and had lasting effects on the nuclear industry in the United States. Approximately two million people in the surrounding region received the equivalent of one-sixth the dose one would get from a normal chest x-ray. Still, the event caused public concern, a loss of public confidence in the nuclear industry, and negative sentiments about the safety of nuclear power as an energy source.

A Nuclear Regulatory Commission fact sheet identifies that in the Three Mile Island incident, either a mechanical or electrical failure prevented the feedwater pumps from sending cooling water to the steam generators that cool the reactor core, resulting in a shutdown of the reactor. The operator opened the relief valve to relieve the pressure, but then the valve became stuck in the open position rather than closing when an acceptable pressure was achieved. The instruments created the misimpression that the valve was closed even as the cooling liquid was being released. No instruments had been installed in the original design to allow for directly monitoring the pressurized water level in the core, so the staff did not understand the plant was "experiencing a loss-of-coolant accident." The issue was compounded as the staff turned off the reactor coolant pumps to control the vibration. Despite the failures, the containment building remained intact and contained the radioactive core meltdown.[8]

Another nuclear disaster, the Chernobyl accident, occurred in 1986 in Ukraine, which was then part of the Soviet Union. The accident has been attributed to a flawed reactor design and inadequately trained personnel. It began as the reactor crew was conducting a routine shutdown to determine how long the power-generation turbines would continue to spin following a loss of power. Prior to conducting the test, the crew disabled the automatic shutdown mechanisms that would have immediately halted the chain reaction in the reactor. Simultaneously, an anomaly in the design of the control rods created a dynamic power surge when they were reinserted into the reactor to halt the chain reaction.[9]

The situation was exacerbated by design and safety flaws, as the interaction of the hot fuel rods and cooling water caused a pressure buildup, resulting in scalding that damaged the reactor and caused a misaligning of the fuel channels for the fuel rods. Shortly thereafter, a severe explosion ensued, which further damaged the reactor integrity and resulted in the release of hot graphite from the reactor core.[10]

Approximately 5 percent of the reactor core burned, with its radiological byproducts released into the atmosphere forming a dangerous radioactive cloud that continued to be emitted for ten days. The result was the death of thirty plant workers within several weeks of the accident. The accident also resulted in the deaths of several hundred others, attributed to radiation-exposure-related illnesses, and the evacuation and resettlement of many of the region's population.[11]

The March 2011 Fukushima nuclear disaster also contains lessons related to the development of technology. A lengthy International Atomic Energy Agency (IAEA) report from August 2015 blames the three core meltdowns on the "blind belief in 'the nuclear security safety myth'" that sees nuclear power plants as so safe that such accidents are "unthinkable."[12]

Legal proceedings were brought against several Tokyo Electric Power Company (TEPCO) executives for shortfalls related to professional negligence. The findings included failure to adhere to international standards regarding operations at nuclear power plants in relation to "periodic safety reviews, re-evaluation of hazards, severe accident management and safety culture."[13]

The response to the disaster was also found to be inadequate to the task at hand, highlighting the need for cultural and institutional reform. In the IAEA report, the highest levels of the government and industry were found to have been at fault. Former prime minister Naoto Kan called the response "confusing" and criticized the company and nuclear regulators for ignoring "legal mandates to implement safety regulations."[14]

Several actions contributed to make the disaster even greater. The failure to remove spent fuel rods and the breach of their holding vessels caused the spent fuel to catch fire, resulting in the release of a radioactive cloud that worsened the contamination. A lack of awareness of the status of the reactors, cooling systems, and peripheral equipment contributed to public panic and questionable decisions being made by TEPCO during the height of the crisis. Public health preparedness—particularly concerning medical countermeasures such as potassium iodide, evacuation of at-risk personnel, collapse of the radiation emergency medical system, and management of long-term psychological issues, to name a few factors—contributed to human suffering resulting from the tragedy.[15]

In each of these nuclear incidents, design flaws in the plants, failure to follow established procedures, and human factors contributed to the magnitude of the disasters. It is also disconcerting that the design flaws were not identified or fully understood until the after-action reviews were being conducted. Interestingly, a 2007 book by Katsunoba Onda, *Tepco: The Darkness of the Empire*, predicted the Fukushima nuclear disaster years earlier.

NOT CONSIDERING THE UNINTENDED CONSEQUENCES

Can you think of any technologies that appear to have been developed with little or no thought to the negative consequences from their use?

In fact, the history of technology suggests that humankind has routinely been slow to introduce laws, policies, regulations, and ethical frameworks, doing so only once the potential dangers of a technology has been demonstrated. The time lag between the first appearance of a technology and the establishment of such control measures normally relates to the natural mismatch between operators using the technology and discovering new use cases versus governments and industry groups that must craft the laws, policies, regulations, and ethical frameworks while not wanting to hinder the legitimate use of the technology. Too much regulation means technology adoption and innovation could be stifled. Not enough and the technology could be misused, with potentially dangerous consequences.

The history of social media provides an interesting example, in which the principal companies do not appear to have considered the unintended consequences of their information-sharing platforms. Even when confronted directly during congressional testimony and in the face of overwhelming evidence, they have been reticent to acknowledge the pivotal—both positive and negative—roles they have played in shaping the early part of the twenty-first century.

No attempt will be made to rehash the rise of social media in this section, as an overview of the history of social media was presented in the chapter on managing technology. Nor will this section attempt to litigate the positives and negatives associated with the social media platforms that have proliferated across the globe and become integral parts of our daily lives.

However, reminding the reader of the beginnings of the internet and its roots as an information-sharing platform seems important to the point that is being offered and remains highly relevant to the type of internet and social media platforms that will exist in the future. If the internet will remain an open architecture platform as was originally envisioned, then less regulation and structure will likely be the result. Alternatively, the internet could become a hybrid that retains a portion that is true to its roots as an information-sharing platform while having another portion that serves as the lifeblood of the global economy, a source of our daily news, and protects our personal information.

How one considers these two alternative internet worlds will set the stage for answers relating to whether international and national laws, policies, regulations, and norms that have served as the foundations for civil society and for managing other technologies will form the basis for the internet of the future. Issues such as net neutrality are directly related to this question. So too are questions of privacy and security of our personal data. To this end, a fundamental question will be whether the future internet will remain useful in the global economy, delivering news and protecting our personal information? Or will an alternative internet that functions as a utility need to be developed?

Terms such as FANG (Facebook, Amazon, Netflix, and Google) and FAAMG (Facebook, Amazon, Apple, Microsoft, and Google) have come to define the major social media players. These companies, some of which did not exist a decade ago, have come to dominate the global financial markets.[16]

However, as we have come to understand, these companies have also been both the targets and purveyors of a range of illegal and questionable behaviors by third parties and as part of their own corporate business models. They have also demonstrated a degree of corporate naiveté regarding the protection of their customers and the need for conducting due diligence in exchange for use of their platforms. It is also clear that little consideration was given in protecting the integrity of their systems—in particular, the veracity of the data—for several of these companies.

After the 2016 election, the degree to which some of the social media platforms were penetrated by Russian hackers became evident. Equally clear was the deliberate misuse of personal data and breaches of privacy by some social media platforms for delivering targeted advertisements and developing "psychographic" profiles of voters. To be sure, vast amounts of data are available, some of which we volunteer through social media posting and others through the digital footprints we leave behind as we go about our daily lives, but understanding and enforcing these limits will be critical to retaining our privacy in this digital age. In some sectors such as health and finance, these limits are well defined and enforced, while in others there is little control over the use of such information. But even in these sectors, our privacy might be on a slippery slope. Consider that if one searches for information on a medical condition, that could be a signal that the user is afflicted with the condition, and this could affect the advertisements the user sees and even be used by third parties such as insurance companies to influence decisions.

In January 2018, Twitter, Facebook, and Google representatives provided testimony to the Senate Committee on Commerce, Science, and Transportation concerning extremist content on their platforms.[17]

This testimony came only after strenuous denials that their platforms were infiltrated and co-opted by Russian hackers likely attempting to influence the US 2016 elections. Generally, the tech companies were criticized for too-narrow focus in their investigations into Russian active measures on their platforms. They also were criticized for under-reporting the scope of the Russian interference.[18]

Furthermore, Congress accused the companies—most notably Facebook—of being "blinded by growth in a way that made them fail to consider the future consequences of building the most colossal, most efficient global information delivery system of the present day."[19]

In identifying the way forward and the responsibility for addressing the issue, one of the senators cautioned, "You bear the responsibility. You've created these platforms and now they're being misused. And you have to be the ones to do something about it. Or we will."[20]

The issue extends beyond the naiveté and questionable activities of social media companies. It also extends into our personal interface with the digital technologies we have allowed into our places of work and homes. Office assistants such as Alexa are constantly turned on, recording all the user's conversations and preferences. This also provides a record that could be subpoenaed and used against the user in a trial. Is this a risk you are willing to take?

TOWARD A RISK-INFORMED TECHNOLOGIST

A fine line exists between overregulating and underregulating technologies. Neither is helpful for technology development. An important factor in determining how much control is required should be based on the anticipated risk the technology is likely to entail. The goal should be to align the regulations with the anticipated technology risk.

Such an assessment can be included during technology development as well as once the technology has been fielded and is in use. For some technology development, the risks are well understood, as are the limits required to engage in safe experimentation. For example, for conducting experiments on diseases, the pathogens are characterized according to their ability to cause disease, the availability of medical countermeasures, and their utility as biological weapons. Based on such criteria, experiments can be conducted at biological safety levels (BSL) that are rated from 1 to 4. BSL-1 is the lowest level, corresponding to a low-risk pathogen, and requires minimal precautions to be taken to protect laboratory workers. BSL-4 is the highest level of containment and is reserved for work on the most dangerous pathogens.

Other technologies are not so well defined. Consider drone technology, which matured essentially in the absence of any formal guidance. The Federal Aviation Administration had little guidance for drones before 2015, when draft drone regulations were posted for public comment. Before 2015, technologists had to rely primarily on their good judgment in deciding what behaviors were acceptable.

While developing a risk-assessment framework is well beyond our scope, having a working knowledge of the components of risk would be helpful for the technologists considering whether a technology should be developed or an experiment conducted.

In the Department of Homeland Security, we considered risk to be a function of threats, vulnerabilities, and consequences. Threats can be naturally occurring or humanmade. For example, both hurricanes and terrorists can present threats to critical infrastructure. Likewise, a design flaw in a bridge could pose a technology-induced threat. A vulnerability describes what could occur—think of the range of occurrences—resulting from a particular threat. Finally, consequences refer to the outcome associated with a particular threat against a known vulnerability.

For the technologist, considering the range of threats, vulnerabilities, and consequences can provide a useful framework for understanding the potential outcomes. By considering the range of risks, the technologist can think beyond the narrow operational envelope in which the technology has been designed to function.

Various analytical techniques can be employed to add more rigor to the risk assessment, should such additional fidelity be judged to be important. The identification of acceptable risk thresholds could also prove to be useful and warranted. In this way, the technologist can assess whether the technology development is being conducted under appropriate controls with proper protective equipment and safety precautions.

THE TECHNOLOGIST'S RESPONSIBILITY

The technologist's responsibility begins with the proper application of the tools and techniques. It also includes providing accurate, reliable, and reproducible research that can serve as foundations for building the broader body of knowledge and aid in technology development. In this way, the technologist becomes part of a larger body of technology and human development that transcends individual projects or studies. The technologist also has the responsibility to reject misuse of technology by colleagues. These are the foundational elements for conducting proper scientific discovery and technology development.[21]

One study developed a set of "ethical principles for scientific research," which included ten principles that could guide the conduct of basic research. The principles were compiled from more than two hundred documents and included duty to society, beneficence, avoidance of conflict of interest, informed consent for human subjects, integrity in research and findings, nondiscrimination of specific groups, nonexploitation of participants, privacy and confidentiality of subjects, professional competence of researchers, and professional discipline.[22]

Beyond these fundamentals, the technologist should embrace the coequal importance of the quest for knowledge and the need for social responsibility in her work. Neither absolute will result in a positive outcome. An absolute quest for knowledge in the absence of social responsibility can result in dangerous and irresponsible behaviors; absolute adherence to social responsibility could imply the technologist avoids taking the prudent risks that are the foundation of scientific discovery and technology development.

Thinking reflectively and considering the risks associated with the technology development should be embedded into the required outputs of the technologist's efforts. Supervisors can reinforce the need for the technologist to report not only on the technical results of the work but also on the policy implications and dual-use potential for their technologies. Asking technologists to consider the second- and third-order effects of their work seems a responsible expectation. And requiring technologists to consider these factors and perhaps even provide recommendations for future uses and managing risks introduces a culture of professional and social responsibility.

CONCLUSIONS

As we have stressed throughout this book, scientific discovery and technological development are not without risk. Managed appropriately and with proper oversight, these risks should be manageable and not inhibit technology development.

This chapter further serves to remind technologists of their broader responsibility to society. To ensure these responsibilities are met, discovery and development must proceed in a systemic manner that considers the established principles and best practices that have come to be expected of the scientific and technical communities.

Scientific discovery and technology development must also always begin from the standpoint that the scientist, technology developer, or engineer has an unequivocal responsibility to society. And finally, there must be the understanding that just because something can be done does not mean it should be done.

CHAPTER 12

REFLECTIONS ON THE FUTURE OF TECHNOLOGY

Having avoided the temptation and remained true to the promise not to develop a lengthy list of technologies to watch for in the future, we are going to provide some thoughts on future technology development. After covering so much ground and addressing such a wide variety of concepts for technology development in the abstract, it seems prudent to discuss the technologies likely to be pivotal in the future.

Let's begin by considering how and why such future technology lists are created. We will then propose some thoughts on the factors that would be important if one were to consider the evolution of technologies in the future, building on the lessons and concepts we have developed throughout this book. Finally, we will identify common trends that are likely to result, regardless of the exact technologies and their development pathways. Doing so would provide a basis for building one's own listing of future technologies.

As we begin this discussion, the reader might be wondering why such technology lists should be developed and what purpose they serve, especially if they arrive at different conclusions. The short answer is that lists have been created to address specific aspects of technology development. Some might be examining specific time frames; others might be focusing on specific technology fields (for example, information technologies); and still others might pertain to examining specific functional areas such as defense and national security applications. In short, they all have a purpose, and the reader should be conscious of this purpose when considering each list. Furthermore, the reader should be aware of the process by which the technology list was developed. Was the list developed by an individual or small group? Does it reflect the results of an industry-wide survey? What factors were considered in the development of the technology list? These considerations are likely to result in differences across lists. With these thoughts in mind, let's look at some of these lists.

DEVELOPING LISTS OF FUTURE TECHNOLOGIES

One 2018 assessment identified "The 30 Technologies of the Next Decade." This list of technologies and their ratings was put together based on five criteria: (1) current and projected size (in terms of direct and indirect revenue), (2) future segment growth rate and scale of adoption, (3) claimed perception of impact on the future (through Wiki-brand surveys), (4) current chatter value and buzz (via Google and social media), and (5) indirect effects and interactions with other technologies and industries.[1] Indeed, these criteria have considerable overlap with the five categories of our technology assessment framework discussed in chapter 6.

However, this list is comparing individual technologies—some transformative and some not—with broader technology fields, which complicates making such comparisons. Ranking future technologies such as artificial intelligence and the Internet of Things on the same scale as collaborative technology and new screens seems both a difficult and an unfair comparison. Perhaps it makes some sense to have a list that compares cats and dogs and mice and elephants as we think about these technologies tactically and over a brief time horizon. Yet this seemingly occurs regularly in many of these lists of future technologies.

In a similar exercise, a 2013 McKinsey study developed a "top 12" list of disruptive technologies using four criteria: (1) the technology is rapidly advancing or experiencing breakthroughs, (2) the potential scope of impact is broad, (3) significant economic value could be affected, and (4) economic impact is potentially disruptive.[2] Even with the smaller numbers of technologies identified, some still seem a bit out of place. For example, identifying advanced oil and gas exploration and recovery with the Internet of Things seems a bit asymmetric, as the former is an extension of current practices, while the latter would potentially be a transformative technology, perhaps even considered a new technology field.

The Gartner Hype Cycle—a method for predicting technology trends—provides another approach for assessing the maturity and adoption of technologies, seeking to identify technologies relevant for "solving real business problems and exploiting new opportunities."[3] The method relies on a graphical approach that considers expectations over time when seeking to place technologies along a continuum consisting of five technology maturity areas: innovation trigger, peak of inflated expectations, trough of disillusionment, slope of enlightenment, and plateau of productivity. Gartner Hype looks at the technologies in increments of less than two years, two to five years, five to ten years, and beyond ten years.

In the 2017 Gartner Hype Cycle, three megatrend areas were considered: AI everywhere, transparently immersive experiences, and digital platforms. Below the megatrend areas, this approach seems to compare more granular technologies. Some are likely to become building blocks and components of larger technologies, such as connect home (also known as smart home), blockchain, machine learning, and edge computing. Others, such as human augmentation, brain-computer interfaces, and neuromorphic hardware, have the potential to be truly transformative and highly disruptive.

In considering such lists, understanding the intended purpose becomes essential. The list of thirty emerging technologies seeks to identify "emerging technologies that will impact culture, the marketplace and society the most over the next decade." The McKinsey list identifies potential disruptive technologies. The Gartner Hype Cycle looks to identify emerging technologies for business and investment purposes.

Therefore, while they use similar criteria and arrive at lists with considerable overlaps, they have very different purposes.

BUILDING YOUR LIST

We should begin by stating the purpose of developing our own distinct emerging technology list.

As has been the purpose throughout our journey, through our list we will seek to understand technology trends with a special focus on societal and security trends that seem likely to be broad and enduring. Undoubtedly the analysis will delve into other potential uses of technology such as investment decisions, despite this not being the designed purpose. In keeping with our previous discussions, the range of the technologies considered would include identifying individual technologies that have the potential to be disruptive and transformative and create discontinuities. Therefore, the analysis seeks to look well into the future, identifying strategic technology trends that are likely to affect humankind. This stands in contrast to the examples listed previously that tended to have more near-term focus.

In building our list of emerging technologies, let's return to the Technology Assessment Methodology. The five categories of the assessment methodology—(1) science and technology maturity; (2) use case, demand, and market forces; (3) resources required; (4) policy, legal, ethical, and regulatory impediments; and (5) technology accessibility—provide a logical point of departure for considering strategic future technology trends.

Three other factors are prescient in considering which technologies are likely to be game changers. The first is the rate of change of a technology. If the rate of change is slow and steady over a lengthy period, the technology could still be worth monitoring and important to the development of humankind, but it is likely not a transformational technology. On the other hand, fields such as information technology (cyber) and biotechnology that are experiencing exponential growth would have to be considered in a different class. The second factor, foundational nature of the technology, relates to the degree to which the technology becomes widespread across numerous applications and even across multiple fields—or, alternatively, understanding whether the new technology is likely to have broader implications for humankind. Consider the importance of polymerase chain reaction (PCR), which has become a foundational element in many advancements in biotechnology, including virtually all applications related to genomics. Finally, the third factor would be the endurance of the technology: will it have longevity or be replaced in months or a couple of years by an incremental improvement?

Perhaps at this point, one might be tempted to develop a formulaic approach that somehow combines these terms—the five areas of the Technology Assessment Model with the three factors listed above—and arrives at a numerical value. However, this would be taking the precision too far. At no time should this analysis entail a quantitative measure of a technology's potential to be transformative. Rather, it should remain a qualitative assessment based on reasoned judgments about the technology's projected availability (as assessed by the Technology Assessment Model) and the transformational capacity (as assessed using the three factors).

Examples can be useful in considering how to apply this method. Think about comparing the transformative nature within military operations of an aircraft carrier versus nuclear weapons. Both have been transformative and contributed to US national security. Yet the use of nuclear weapons instantaneously ushered in the age of nuclear weapons, was highly disruptive, and fundamentally changed the concepts of strategic deterrence and warfighting. In relations between major powers, one can essentially talk about the Prenuclear and Nuclear Ages. While critically important to projection of US power and operational control of the maritime domain, the same cannot be said of the aircraft carrier. Using this comparative function, judgments can also be made comparing technologies in different fields—from, say, advanced energy, materials, and healthcare—to determine the degree to which they are likely to be transformative.

The problem this transformative technology function is trying to prevent is the development of a list that ranks future technologies such as artificial intelligence and the Internet of Things on the same rating scale as collaborative technology and new screens.

So, if the desire is to separate the technologies that are likely to cause disruptive and discontinuous shifts from those resulting in small and incremental shifts, then an approach such as the one recommended seems prudent. Stated more directly, comparing the invention of the printing press with new screen technology seems trivial and like not much of a comparison. The former caused an inflection in the history of humankind; the latter would likely only provide enhanced capability to an existing system.

This is not meant to denigrate the importance of incremental advances in technology, as over time they combine to create important, measurable benefits across a range of technologies, as our aviation example in chapter 3 demonstrated. The recursive nature of technology will incorporate combinations of incremental and elemental technologies that become part of larger, more transformative systems and technologies.

And if this is the case, perhaps these technologies should be compared with other elemental technologies, while seeking to identify those individual technologies and technology fields likely to be truly transformative.

Perhaps the reader is wondering at this point if this discussion of lists even matters and what it might lead to. In fact, it does matter, as, armed with concepts for how future technologies are likely to evolve and thus the trajectory for humankind, we can perhaps shape the future in ways to be friendly (or at least not hostile) to humankind and the environment.

Previously, we introduced the potential for the convergence of biotechnology, AI, and the IoT in ways that could fundamentally alter the essence of what it means to be human. However, this convergent future state was not meant to be predictive, nor is it a foregone conclusion. Humans can steer the development paths now that technology will take in the future.

And it begins with developing lists of technologies to watch—perhaps those that could be disruptive and transformative, either individually or in combination. For some technologies, developmental factors and availability of a technology will combine to create a tipping point where increased monitoring or, in some cases, limiting research, activities, and proliferation of a technology could prove prudent. Such control measures could include voluntary compliance, codes of ethics, institutional review boards, policies, regulations, laws, and international agreements.

However, implementing control measures—or attempts to manage technology—could also have negative effects by limiting the development of an important technology with potentially great societal benefit. The question of technology limits must also be considered within the context of an increasingly globalized research and development environment. In such an environment, limits on technology development would invariably be implemented differently across the globe. Of course, if some nations do not abide by the limits or interpret them differently, asymmetries and strategic advantages and disadvantages could result. To prevent such imbalances and even strategic surprises, nations would do well to regularly assess technology risks to their national security, economic competitiveness, and societal well-being. If technology threats and risks are discovered, nations can adjust their priorities accordingly.

Throughout, the goal should be to maintain the delicate balance between greater technological advancement and the safety and security of society. Development of a risk-based framework for guiding efforts to assess when additional scrutiny or even control measures could be required would be prudent, along with an assessment of which measures might be most effective.

Separating the technology cats and dogs and mice and elephants would also be a worthwhile endeavor, as it could assist in focusing attention on those areas most likely to be transformative and assist in better shaping the future codevelopment of humankind and technology.

COMMON EMERGING TRENDS IN THE FUTURE RESULTING FROM TECHNOLOGY DEVELOPMENT

As Yogi Berra, the colorful former Major League Baseball player, coach, and commentator, once quipped, "It's hard to make predictions, especially about the future." And the same can be said about making technology predictions. Yet five trends are emerging that will shape all future technology development.

Trend 1. Continued Shifts in Technology Development

Many more people are engaged in S&T, R&D, and I&T activities than at any time in human history. They are conducting more experiments, spending more money, and achieving more technology breakthroughs than any time previously. Global trends are shifting, as industry will continue to dominate in technology development activities. Economic incentives will drive these tech development forces. Government will need to focus its technology development activities in those specific fields that are not being serviced by industry. The government (especially the military, intelligence, and law enforcement communities) will need to become better consumers to incorporate cutting-edge nongovernment-sponsored capabilities into their systems and processes.

Trend 2. The Global Economy Makes Managing Technology More Challenging

The numbers and combinations of emerging technologies will inhibit the development of prudent control measures to protect proprietary technologies, limit dangerous activities, and promote technological advancement. Identifying strategies that balance the need for continued technological advancement with that of the safety and security of society will be challenging. Technological surprises will continue to occur, and new disruptive technologies are likely to emerge. Technology proliferation will continue long before the implications are well understood. Governments and nongovernmental stakeholders will be challenged to proactively establish control measures such as codes of ethics, standards, regulations, policies, and laws—this will be especially of concern for emerging technologies and combinations of technologies (such as the biotechnology, AI, and IoT example from chapter 10). The convergence of technologies will continue to lead to greater numbers of unexpected use cases.

Trend 3. Increasing Risk to Humanity

Humankind is developing more powerful technologies, with increasing potential to both positively and negatively affect society. Technologists are working across different fields, fueling advancements and even developing new technology fields. And the stakes will only become higher as technologies converge and continue to be combined. We should also expect that the number and frequency of emerging disruptive technologies will continue to increase. These advancements portend potentially great societal benefit but also increased risk as more powerful technologies become more readily available. Technologies once requiring great skills and education have become mechanical processes and are becoming more readily available to even lower-skilled individuals. The chances of deliberate misuse or accidents will increase. The potential for statelike capabilities in the hands of nonstate actors, even lone actors, continues to increase. Timescales between technology-fueled events will decrease, yet the magnitudes of such events will increase. While technology might not always be the cause of an event, it will undoubtedly serve as an accelerant.

Trend 4. Growing "Competition" between Humans and Computers

Throughout history, we have witnessed changes in patterns of the workforce based on the infusion of technology. Initially, technology allowed for more efficient processes and the substitution of technology for labor, such as in agriculture. With the widespread use of robotics in manufacturing, robots were substituted for human labor in repetitive, mechanical, and sometimes dangerous processes, resulting in greater effectiveness and safety at less cost. These changes caused a reshaping of the workforce away from agriculture and manufacturing and toward services. Through the combination of robotics, autonomous systems, and artificial intelligence, further opportunities exist to substitute non-human systems into areas that have been traditionally the purview of humans. Such trends could result in displacement of the workforce and the need for different, more technical workforce skills, but there will also be a societal cost. The growing use of AI for security and defense is creating significant ethical and moral issues that will need to be addressed. Do we really want to allow warfare to devolve into a clash of robots, or should it remain a fundamentally human endeavor? In these emerging areas where nonhuman systems displace humans, tensions will arise, and great care will be required to navigate the future. Attention needs to be paid to establishing responsible limits on acceptable substitutions. Human-machine teaming and collaboration must become a focus to better account for the strengths and weaknesses of each.

Trend 5. Changing Expectations of Privacy

The conception of privacy as guaranteed in the US Constitution and Bill of Rights has been eroded since the mass infusion of information technologies. Through devices that track our health, movements, and social media status, we have changed the very nature of privacy. The technologies developed to improve our daily lives now intrude on them. In some cases, these intrusions have been inserted in our lives without our consent, such as cameras mounted on city streets. In other cases, we have invited them in through social media platforms such as LinkedIn and Facebook. We will need to redefine privacy in light of IoT, big data, and social media advances. With increasing use of biometrics and greater ability to correlate information about individuals and society, the limits on encroachment on privacy will increasingly continue to be tested. Our growing demand for situational awareness to manage our daily lives will also clash with personal privacy. In a grand bargain, we will continue to allow social media and passive and active electronic surveillance into our lives, despite the erosion of privacy they bring.

CONCLUSIONS

We have remained true to the promise to not develop a list of technologies, but we have examined what others have said in evaluating technologies. Many of the lists have been created to examine specific topics, covering specific time periods. But our purpose was to provide a structured way to identify transformational technologies likely to affect individuals and society in fundamental ways.

The five trends that have been highlighted in this chapter do not describe the future of humanity and technology, but rather highlight the forces that are likely to be present as we look to the future. Furthermore, they are not intended to be deterministic. A clash of robots on a future battlefield does not have to be the reality any more than a society without privacy has been predetermined. We have the opportunity to intervene, and now is the time to do so.

CONCLUSIONS

In many respects, the science and technology foundations of today are based on Vannevar Bush's report *Science, The Endless Frontier*, a document written more than sixty-five years ago. However, this does not mean that the future history of technology has been written. Doing so would require reexamining the unbreakable link between humankind and technology, understanding the forces that have shaped this codevelopment, and influencing the collective future.

One cannot help but wonder how Vannevar Bush would view the evolution of science and technology stemming from this single report. Would he recognize the national "S&T structure" as derivative of his ideas? Would he see the current structure as allowing us to take advantage of the opportunities while mitigating the challenges? Would he believe research has been appropriately resourced? Or has it been frozen out by the requirements of the here and now?

The Story of Technology proposes thinking differently about technology, suggesting new ways to assess how technologies are evolving and are likely to evolve in the future. This book began by defining technology as the purposeful application of capabilities (including scientific knowledge or the collection of techniques, methods, or processes) for practical purposes. As we reflect on the future of technology, it is useful to come back to this definition.

Throughout the course of human development, technology has been a fixture. This is unlikely to change and therefore must be shaped so that the technologically enabled world of the future is hospitable to the human experience. This outcome is not assured and requires constant addressing to safeguard the tenuous balance between discovery and prudence.

We have witnessed waves of technology-enabled globalization, driven by advances in communications, transportation, and information technologies. These waves have contributed to an interconnected world that can be traversed in a matter of hours; where we know more about each other than at any point in history; where regional economies, societies, and cultures have been integrated; and where technology-enabled international institutions have been developed to promote peace and stability. We have also witnessed the dark side of globalization, in which the mixing of peoples and cultures has led to instability and strife, competition for resources, and the rapid spread of disease, to name a few.

In the early sections of this book, we spent considerable attention working to deconflict the terms *science and technology*; *research and development*; and *innovation and transformation*; certainly, this was important for grounding the remainder of our discussion. However, in practice, delineating the differences with great precision is not possible. Nor is it that important to do so, for wherever these sorts of activities are occurring, one will find some combination of scientists, engineers, technologists, innovators, and operators working to advance scientific discovery and technology development.

CONCLUSIONS: A GLIMPSE OF THE FUTURE OF TECHNOLOGY AND SOCIETY

This book has been a journey about the relationship between humankind and technology. The purposeful application of capabilities to solve practical problems has led to this symbiotic relationship.

From the earliest days, humans have sought ways and means to survive and become self-sufficient. In the same way that humans have evolved over millions of years with our DNA recording this journey, so too has technology. The early building blocks of technology can be seen in even the most sophisticated of modern technologies.

Throughout, we have also witnessed that the evolution of technology is not linear. Each technology matures at its own pace. Sometimes a technology requires additional science or technological development for its component technologies to reach maturity. At other times, technology perhaps even leads humankind, as new discoveries lead to new applications for use. Consider penicillin, which was discovered in the soil and became the miracle drug of the twentieth century.

Unfortunately, history has also provided some important warnings that should be considered as we look to the future of technology. The converging technologies of biotech, artificial intelligence, and the Internet of Things have raised the stakes for humankind. Taken individually or collectively, each has the potential to change the human experience in very fundamental ways, perhaps redefining our physical, mental, and emotional states of being.

Yet these are not the only technologies or combinations of technologies that have this potential. It is for this reason that we must chase the future of technology to allow us to better understand the opportunities and challenges that might lie ahead.

SOME FINAL QUESTIONS FOR CONSIDERATION

Future interactions between society, technology, and the potential for increasing globalization lead to numerous questions. Many of the issues implied within the questions have been discussed at length throughout the book, while others have been directly implied. By listing these questions in a single location, the goal is to provide the reader with an opportunity for final contemplation of these issues. They are at the heart of the question of the future of technology.

Maximizing Technology's Benefits

- What is the proper balance between connectedness and disconnectedness in science and technology development in the future?
- What strategies can be undertaken to promote technology development now and in the future?
- Is there a proper balance to be struck between overregulating and underregulating technologies?
- Can technology centers, smart cities, and innovation hubs stimulate and promote technology development?

- What are the technologist's responsibilities in preventing, mitigating, or responding to issues resulting from the use (or misuse) of their products?

Diffusion of Technology

- As technology diffusion continues, do traditional methods of managing technology become irrelevant, and do new methods need to be identified?
- Will the internet become untrustworthy or even unusable? And if so, given the oversize impact of Information Age technologies, how will this affect technological development?
- Will technology proliferation enable subnations to gain de facto nation status?
- In many cases, policies for managing technology focus on a single technology area, but what if the potential for harm is not the individual technology but the combination of multiple technologies?
- Has the world witnessed the use of weaponized social media? Is the weaponization of the internet appropriate?

Resourcing Technology Development

- Given the diffusion of technology, what should be the strategies for ensuring the United States maintains technological superiority?
- Should current trends in government versus industry spending on R&D be consciously adjusted? If so, how should they be adjusted?
- What are the cultural elements most important to science and technology, research and development, and innovation and transformation?
- Should the results of US government-sponsored R&D be shared globally?

Technology and Ethics

- Will it be possible to halt the changing of what it means to be human?
- Is some research and development too dangerous to be conducted under any circumstances?
- Are there boundaries or limits that technologies should not be allowed to exceed? If so, are there practical ways to limit these behaviors?
- What is the responsibility of technology developers to notify states of potentially dangerous activities that are occurring within their borders?
- With many technologies having implications for individual privacy, what should the expectations be for humankind in the future?

Effects on Society and Relations between Nations

- What will changes in technology mean for relations between nations and between the sovereign and the governed within nations?
- If technologies can facilitate or accelerate events such as the Arab Spring, what role can they play in mitigating the effects of such events?
- What is the effect of technology on national borders? Correspondingly, given the proliferation of technology, is it reasonable to expect states to be able to control technologies within their borders?
- If the combination of AI, robotics, and autonomous systems eliminates 1.2 billion jobs and $14.6 trillion in wages, what are the implications for society? Do companies have a social responsibility to mitigate these effects? What changes to workforce training and education will be required, given the anticipated changes resulting from automation?

- What are the implications of the use of autonomous weapon systems on the future battlefield that include the entire kill chain from observation and detection to the lethal use of force?

APPENDIX A

THE TECHNOLOGY LEXICON

In chapter 2, we provided brief context and definitions for the key terms, but such brevity by necessity omits much of the detail. In this appendix, the terms *science and technology*, *research and development*, and *innovation and transformation* are explained further.

Science

Science is the quest for fundamental understanding of natural phenomena.

While separating the terms *science* and *technology* over the course of human history is difficult, even identifying with any confidence when science began is challenging. The introduction of the written word and mathematics was certainly key to the development of the scientific method. Without them, making observations, taking measurements, and conducting repeatable experiments were not possible. This then implies that the scientific method, which we have included in our definition, could be no earlier than 3000 BCE, when evidence of the first peoples to use math—the Babylonians and Egyptians—was identified.

The derivation of the word *science* provides some clarity but little distinction from the concept of technology. Science has a long etymological history, coming from the Greek word meaning "true sense"; from the Latin *scientia*, which means "knowledge"; and from Old French meaning "knowledge, learning, application."[1]

Still, charting the history of science is not an "exact science." Medicine, which is certainly considered a science, can trace its roots to ancient Egypt. But some of the elements of the scientific method were not incorporated into the Egyptian practice of medicine.

Likewise, Aristotle's principles of reasoning and logic begin to provide necessary components of the scientific method. He was conversant in most scientific topics and certainly contributed to the development of scientific thinking, as his deductive reasoning was and continues to be important to the development of the scientific method.[2] In the same vein, Pliny the Elder—who was born Gaius Plinius Secundus—was a Roman author, naturalist, and philosopher, and a naval and army commander of the early Roman Empire who lived from 23–79 CE. He is credited with developing the first encyclopedia, a thirty-seven-volume set of work touching on a variety of subjects including astronomy, geography, botany, geology, medicine, and mineralogy, to name a few topics. His works certainly cover a great deal of ground, yet they reflect observations across specific areas rather than use of the scientific method in examining these topics.[3]

In examining this question of when the scientific method began, two points in time and areas of the world stand out as being likely worthy of the distinction. The first are Muslim scholars between the tenth and fourteenth centuries. One prominent scholar was al-Haytham, who has been credited by some as the "architect" of the scientific method composed of the following seven stages:[4]

1. Observation of the natural world
2. Stating a definite problem
3. Formulating a robust hypothesis
4. Test the hypothesis through experimentation
5. Assess and analyze the results
6. Interpret the data and draw conclusions
7. Publish the findings

Al-Haytham and those from the Muslim cohort are largely credited with influencing the development of the scientific method during the Renaissance. Roger Bacon (1214–1284) contributed the idea of inductive reasoning, or the use of specific observations to support the probability of a general conclusion. He also contributed an updated version of a model for the scientific method based on the Islamic model. In this model, the four steps were observation, hypothesis, experiment, and verification.[5]

During the Enlightenment, Francis Bacon (1561–1626) and Descartes (1596–1650) sought models for the scientific method using inductive or cause-and-effect models and deductive reasoning, respectively, in their development efforts. Later, Sir Isaac Newton promoted the idea that both induction and deduction must be part of the scientific method.[6]

Today, the meaning of the term *scientific method* varies across areas of study yet consistently encompasses some basic elements:

1. Formulate hypothesis
2. Collect data
3. Test hypothesis
4. Conclude

Koch's postulates provide a practical example of the scientific method. Robert Koch was a nineteenth-century scientist who was studying the relationship between bacterial pathogens and disease. His work included the study of anthrax and tuberculosis. In examining the diseased tissues of animals that had succumbed to anthrax, he observed that the blood of animals was infected with rod-shaped bacteria. Through this observation, Koch developed an experimental design that identified a possible link between the bacteria and the disease. The four steps of Koch's postulates were to (1) observe signs of causal organism and symptoms of disease, (2) isolate the causal organism in a pure culture, (3) inoculate a healthy host with the suspected pathogen, and (4) determine if this organism, when inoculated into the host, caused an identical disease.[7]

In describing science, one could begin with a set of principles. Several would come to mind. Science should be conducted using a process or method that is the same for each experiment. It should be reproducible: an experiment employing the same inputs and methods should provide the same output. Inputs and outputs should both be measurable. Measuring inputs ensures known quantities are used; measuring outputs allows for accurate results. Science should also include the development of conclusions, often in the form of principles, as is common in the scientific method.

One definition states that science is "the pursuit and application of knowledge and understanding of the natural and social world following a systematic methodology based on evidence."[8] It goes on to identify several tools of scientific methodology:

- Objective observation: Measurement and data (possibly although not necessarily using mathematics as a tool)
- Evidence
- Experiment and/or observation as benchmarks for testing hypotheses
- Induction: reasoning to establish general rules or conclusions drawn from facts or examples
- Repetition
- Critical analysis
- Verification and testing: critical exposure to scrutiny, peer review and assessment

Another definition defines science as "the intellectual and practical activity encompassing the systematic study of the structure and behavior of the physical and natural world through observation and experiment."[9] Still another describes science as "the field of study concerned with discovering and describing the world around us by observing and experimenting. Biology, chemistry, and physics are all branches of science."[10] This final definition elaborates on the definition by adding,

> Science is an "empirical" field, that is, it develops a body of knowledge by observing things and performing experiments. The meticu-

lous process of gathering and analyzing data is called the "scientific method," and we sometimes use science to describe the knowledge we already have. Science is also what's involved in the performance of something complicated: "the science of making a perfect soufflé."

As one can see, arriving at a single definition is not without challenges. The Science Council spent over a year looking at the definition of science and concluded that "science is the pursuit of knowledge and understanding of the natural and social world following a systematic methodology based on evidence."[11]

For our purposes going forward, it is important to have an agreed definition of science. Our definition draws on the yearlong Science Council effort but will introduce the rigor associated with scientific discovery. Our resulting definition of science will be:

Science is the pursuit of knowledge and understanding of the natural and social world following a systematic (or scientific) methodology based on evidence derived from observations, experimentation, analysis, and repetition of results.

The systematic (or scientific) methodology contained within our definition will be shortened for our purposes to the term *scientific method* and will be further defined as the formulation of a hypothesis, collection of data, testing of the hypothesis, and drawing of conclusions.[12]

Regardless of the precise definition, it will suffice to stipulate that science is based on the use of the scientific method that provides a means of understanding a phenomenon, organizing the body of thought surrounding how it works, and drawing conclusions and principles regarding an area of study.

Technology

Technology is a word that is often used yet largely remains undefined within the casual conversations that occur on the topic. In fact, the word was not commonly used until roughly the mid-twentieth century.[13] Even for those who are looking to do a serious examination of some aspect of technology, few rigorously define the term.

Earlier, we have talked about the codevelopment of humankind and technology. The lack of distinctions between the activities meant that early science and technology could be placed within the overarching category of technology. This shorthand was judged to be acceptable, as early humankind used technology for practical purposes rather than specifically employing a scientific method such as that contained within our definition of science. This is not to say that early humans were not concerned with repeatable technologies. They undoubtedly desired to develop tools, utensils, weapons, and hunting implements that achieved the same outcome each time they were employed. And the technologies that were developed were clearly based on—or at least had some of the elements of—observation, experimentation, analysis, and repetition of results, as captured within our definition. However, such efforts did not directly seek to develop a body of knowledge, to systematize, or to develop facts and principles, all of which today we associate with scientific discovery.

We have described technology as being about solving problems. This was a shorthand designed to allow us the flexibility to examine the early science and technology without having to make a clean delineation between the two. However, as with the definition of science, definitions of technology vary greatly depending on their origins.

One source defines technology as "science or knowledge put into practical use to solve problems or invent useful tools."[14] Such a formulation implies that science or knowledge in all cases precedes technology. But is this always the case?

Another definition identifies technology as the "purposeful application of information in the design, production and utilization of goods and services, and in the organization of human activities."[15] This definition tends to broaden the idea of technology beyond a tool, allowing for applications in processes, services, and organizational activities. It also omits the idea of knowledge and science as inputs into the development of technologies. Such an omission is made more interesting because this definition comes from a business perspective rather than a scientific one.

Another source defines technology as the "methods, systems, and devices which are the result of scientific knowledge being used for practical purposes."[16] Once again, this definition implies that a meaning of technology beyond science or development of tools will likely be important. It too allows for "methods," which implies adjusting techniques to solve problems for practical purposes.

Clues to the meaning of *technology* can be found in its derivation, from a Greek word that is often translated as "craftsmanship," "craft," or "art," implying that it goes far beyond simple toolmaking or into the skill with which those tools were to be applied.

By now, it should be clear that different fields of technology have adapted the basic premise of technology to suit specific disciplines. However, for our purposes going forward, the concept of technology will be applied in the broader sense, capturing the roots of the term that come from its derivation, bringing in the idea that technology is broader than science and knowledge. Therefore, our definition will be:

Technology is the application of capabilities for practical purposes. It can include inputs from scientific knowledge or the collection of techniques, methods, or processes applied in a purposeful way. The term applies across a broad range of human endeavors.

Given the definitions of science and technology—both the definitions of others and our derived ones—it is not surprising that the terms have frequently been used interchangeably and often as a pair.

Research and Development

Research and development, normally referred to simply as R&D, is a relatively new concept beginning in the early part of the twentieth century. For those working in the field, the terms *research and development* and *science and technology* at times appear to be interchangeable. *Research* seemingly aligns very nicely with *science*, while *development* and *technology* appear to have a similar comfortable alignment. However, such a simple translation of the terms would fail to recognize important distinctions that exist between them.

To demonstrate the symbiotic relationship between these two

pairs of terms, the definitions of technology readiness levels provide an important starting point. TRLs provide a systematic approach to identifying the maturity of various technologies. Appendix B provides definitions for the nine technology readiness levels.

Another important distinction between science and technology and research and development relates to the intended purpose of the actions. Science and technology have been described, respectively, as activities undertaken for gaining a fundamental understanding of natural phenomena and as using scientific knowledge for practical applications.

In contrast, research and development are clearly related to activity designed to result in attainment of new capabilities for a specific purpose. This also closely relates to activities associated with an industrial process, procurement, or acquisition. In fact, the term R&D derives from the Industrial Revolution, when companies engaged in research with the intent of building on scientific principles and ultimately applying their findings to developmental activities that would result in products that could be bought and sold; in other words, their focus was on moving products through an acquisition life cycle.

Before continuing this discussion, it is imperative to once again stress that science and technology and research and development are not linear processes. Rather, technologies are eventually developed through application of the scientific method, experimentation, learning, and failing. In other words, one should not expect to go through the process linearly without having to return to earlier stages. Perhaps more needs to be understood about the science or the technological principles before moving forward. Or perhaps new technologies need to be developed to continue to make advances. The nature of discovery is that as more is learned, new questions arise, generating additional requirements for scientific discovery, additional research, and further development before being able to field mature systems.

Government and industry have come to rely on technology readiness levels to provide a common terminology to indicate how a technology has progressed in its life cycle. In a sense, these TRLs have become a shorthand that allows for objective measures of scientific and technological maturity. This structured framework enables one to answer important questions regarding a technology or a system that is being developed. Are fundamental principles still being identified, or is it fully matured and in use? Are the basic principles understood? What is the risk associated with the technology development? Is it in the preliminary stages, when technology development normally requires significant funding to gain modest improvements? Or has it reached a maturity whereby it has become a component of other emerging technologies?

R&D employs technology readiness levels defined from 1 to 9 and divided into three basic categories: research, development, and life cycle fielding.[17] Research builds on the scientific foundations and is divided into basic and applied, corresponding to TRLs 1–3. Basic research calls for observing and reporting basic principles in the same way as science does. Formulating a hypothesis, collecting data, testing the hypothesis, and drawing conclusions are all important to basic research, just as they are to scientific discovery. Applied research differs from basic research in that one has a use case in mind. In the applied research, a concept for the use of the technology is developed, and a potential application is formed.

One important distinction should be highlighted regarding the comparison between basic research as part of scientific discovery versus R&D. For scientific discovery, one seeks to gain a fundamental understanding or knowledge about a natural phenomenon. In contrast, for basic research conducted as part of R&D, there is likely some overarching concept of use. It is difficult to envision that corporations today would engage in basic research with no expectation of a return on their investment. Later in our discussions of scientific discovery, we will highlight the essential role of government in promoting basic research.

Development entails a broad range of activities designed to mature the technology prior to fielding and use in an operational environment, or TRLs 4–7. The dividing line between research and early technology development really occurs between TRLs 3 and 4. The subcomponents of development include developing breadboards of the components and testing them in a laboratory environment, validating them in a relevant environment, incorporating them into systems, demonstrating these in a relevant operational environment, and finally developing a full prototype that can be tested in an operational environment. The final two steps in the TRL process, TRLs 8 and 9, are actual system tests and demonstrating the technology, and proving the technology through successful mission operations, respectively.

It is important to note that no universal definitions of *research* and *development* exist. For example, each cabinet-level department and agency of the US government has a different definition tailored to meet its specific R&D process. While they have the same purpose—providing a structured way to discuss technological maturity—and do not vary greatly, important subtle differences do exist. For example, the Department of Homeland Security has definitions of these terms that even more closely align with science and technology as presented above.[18]

Yet the Department of Health and Human Services contains TRL definitions that correspond to the R&D processes associated with acquisition of therapeutics and diagnostics. Their eight categories are

1. review of scientific knowledge base
2. development of hypotheses and experimental designs
3. identification and characterization of preliminary product
4. optimization and demonstration of activity and efficacy
5. advanced characterization of product and initiation of manufacturing
6. regulated production, regulatory submission, and clinical data
7. scale-up, initiation of GMP [good manufacturing process] process validation, and phase 2 clinical trial(s)
8. completion of GMP validation and consistency lot manufacturing, clinical trials phase 3, and FDA approval or licensure.[19]

No attempt has been made to provide an exact cross-walk of the terms S&T and R&D. As the above discussion indicates, close relationships exist between science and research and between technology and development. In the case of science and research, principles are being developed, leading to a better understanding of a natural phenomenon. For technology and development, the goal is practical uses.

Innovation

Innovation is yet another term that frequently gets interwoven with science and technology and/or research and development. However, innovation has a very different connotation, relating to the novel use of an existing capability. One definition states that innovation is "the process of translating an idea or invention into a good or service that creates value or for which customers will pay."[20]

In contrast to S&T and R&D, which are methodical processes, innovation relates more to the generation of ideas and the development of combinations of technologies and capabilities in a far less structured approach. In describing innovation, innovators often talk of developing ideas for solving problems or improving efficiency. Innovations result from whittling down the list of a wide variety of ideas to a small subset or even a single innovation that will ultimately be pursued. In this way, perhaps only one or two original ideas would eventually be developed into something (i.e., a good or service) for which a customer would pay.

Innovation can be evolutionary and reflect a continuous process of change and adaptation to retain current customers and attract new customers. One often sees evolutionary innovation in products that are updated annually, such as automobiles. From year to year, small updates will be incorporated that improve functionality or safety, yet the basic model remains largely unchanged.

Innovation can also be revolutionary, or discontinuous, where a new product is introduced into the market that captivates the interest of customers.[21] The introduction of smartphones, which provided mobile mini-computers and communications, certainly qualifies as a revolutionary technology. However, Apple's line of iPhones with updates spaced every couple of years, introducing new products to a committed customer base, can be thought of as evolutionary.

Likewise, the recent Fitbit craze represents another revolutionary or discontinuous innovation. This new type of device attracts people at the intersection of those concerned with monitoring health, having convenient electronics at the ready, and keeping a record of their daily moves.

The rise of companies purveying social media demonstrates another type of innovation, bringing "products" that have not existed previously to the marketplace. It represents a discontinuous innovation that is nearly unrecognizable from its predecessors of newspapers, the news media, and other forms of information sharing.

No endeavor has seemingly been insulated from these dramatic changes. Our personal communications with friends and family, professional relationships, and financial dealings have all been changed through the use of social media and other information- and communications-technology capabilities. Even military operations and intelligence collection have evolved as social media has become an important tool in targeting adversaries and gaining an information advantage. For example, the use of open-source intelligence (including social media, web searches and traditional news sources) in use by the Intelligence Community has continued to expand as a proportion of the source of intelligence gathering in recent years.

As the examples have demonstrated, the initial innovations are often considered revolutionary but, over time, transition to evolutionary innovations. Some might even consider these later innovations to be necessary corrections to products that were rushed to market before all the engineering was completed. In this way, one might even consider some innovations to be more like prototypes than final products. For "hot" products, customers are often willing to overlook some degree of "clunkiness" to have a product that is first to market and fills a perceived important need.

In many respects, innovation is less a formal process for developing and improving products than an ability to tap into the psyche of the market. Doing so provides an opportunity to develop and harness promising ideas that will captivate the market and for which people will be willing to pay. Given such a construct, it should come as no surprise that companies that are highly innovative and able to adapt likely have a culture that encourages, promotes, and rewards innovative ideas. Innovation is not just about products, either. It is possible to innovate processes and gain greater efficiencies through the implementation of innovative solutions.

Today, the concept of innovation centers has taken on a life of its own. For example, Accenture has an innovation center in Boston that is designed to rapidly prototype and deliver software to customers. The Department of Defense has established innovation centers in Silicon Valley, Boston, and Austin to try to rapidly identify emerging trends that have the potential to be used for military applications. The Defense Innovation Unit Experimental (DIUx) was established for the express purpose of developing technology solutions with greater public-private cooperation for defense customers.

According to former secretary of defense Ash Carter, who established the DIUx shortly after taking over as secretary, the nature of the problem was that "commercial industries would soon overtake defense labs at innovation and that to protect US global interest, the Defense Department would need to form new relationships with the private sector." Despite the best of intentions, one account lamented, "There remained the main challenge: how to cut through the Pentagon's Byzantine procurement process."[22]

Seeking to increase effectiveness and efficiency has led government, industry, and even individual consumers to pursue innovation as a top priority. In some respects, this pursuit feels more like an effort to short-circuit the S&T and R&D processes, providing access to "innovation" without the heavy investments required of early S&T and R&D. Whether such a strategy will be effective in the long run or as a substitute for traditional S&T and R&D remains an unanswered question.

Transformation

Transformation is a term that describes a comprehensive change within an organization. The process of transformation can occur in any organization, industry, or government. However, the term *transformation* owes a significant portion of its popularity to its use (some might say overuse) by the US military.

Transformation was initially used in military circles to describe a very comprehensive change that affects nearly all aspects of the organization being transformed. This is not to say that government and civilian organizations were not undertaking meaningful change previously but, rather, that the military was instrumental in coining this use of the term.

There are certainly examples of businesses that have undertaken transformations to improve their organizational structure, processes, and outputs. For example, the MITRE Corporation proposes a framework for organizational change management that includes requirements for leadership, communications and stakeholder engagement, knowledge management, enterprise organizational alignment, and site-level workforce transition.[23] The change envisioned through this organizational redesign process was no less significant than that envisioned by the military beginning in the late 1970s.

Indeed, there are examples such as Toyota, which has created fundamental change throughout the organization through business process transformations. The Lean Six Sigma approach described previously actually fits nicely as both an innovation and a transformation. Often information technology professionals and companies employ business transformation in updating their enterprise hardware, software, and processes. By updating all three of these areas, one seeks to avoid updating a bad (or obsolete) process with new hardware and software.

Corporate transformations can be more fundamental as well. The restructuring of the large integrators such as L-3, Boeing, and General Dynamics, which have spun off or consolidated their services sectors, represents a significant transformation of their organizations and the products and services they offer. In the case of L-3, the company divested of almost 33 percent of the company's revenue when the services were divested, eliminating almost $5 billion worth of business.

Other examples of significant transformations include IBM, which changed its product line from commercial scales and punch-card tabulators to massive mainframe computers and calculators and now to selling software, consulting services, and IT services. Similarly, the Xerox company originally manufactured and sold photographic paper, yet today it sells toner, ink, software, scanners, inkjet printers, and copy machines. Monsanto, founded in 1901, had as its first product saccharin and now sells herbicides and genetically engineered seeds. General Electric, which was founded in 1892, sold electric inventions including the light bulb yet now sells gas engines, hybrid locomotives, high-definition CT scanners, ultrasound devices, and chemical sensors.[24] Each of these examples represents a significant transformation to account for shifts in its market, reaction to these changes, and introduction of new offerings.

The National Security Act of 1947 transformed the national security foundations of the United States. It established the US Department of Defense, the Central Intelligence Agency, the National Security Council, the position of chairman of the Joint Chiefs of Staff, and the Department of the Air Force. This legislation developed in the aftermath of World War II was based on perceived requirements by the United States for a changed role in global affairs. The legislation has served as the central organizing construct for the US national security apparatus ever since.[25]

These sorts of transformations can have second- and third-order effects as well. For example, the Department of Defense has been under nearly continuous transformation since its inception with the National Security Act of 1947. Over the nearly seventy-five years since the act was signed, the military has continued to transform to meet the changing strategic environment and the threats, challenges, and opportunities that have been presented. As a result, changes have been made to organizations, weapons, and doctrine and have dramatically altered the way the force fights.

Throughout military history, armies have continued to evolve, from the antiquities to Napoleonic warfare to Industrial Age warfare and now to Information Age warfare. During the Cold War, a major transformation effort—the Revolution in Military Affairs (RMA)—was first introduced by the Soviet Union in the 1970s and 1980s and later adapted by the US military. Interestingly, the Soviet Union referred to this effort as the military technological revolution. Marshall Nikolai Ogarkov, the head of the Soviet Armed Forces, theorized that innovative technologies were on the horizon and would need to be integrated into the force. They included such systems as directed energy weapons and capabilities for space control.

The US military, largely in response to the Soviet initiative, developed its own version of RMA that was centered around the incorporation of digital technologies into the force for applications from warfighting to logistics to command and control. Operation Desert Storm in 1991 served as a demonstration of the potential of both the power of the digital force and the need to continue the transformation. The US military underwent a series of RMA initiatives centered around network-focused warfare that linked all aspects of the force into a digitally enabled military. Key concepts included advanced weapons technology, dominant maneuver, precision fires, intelligence, surveillance, reconnaissance, and focused logistics that would enable the force to project power and be dominant when it arrived.

The 1995 report *Joint Vision 2010* became the operational description of how the US joint forces intended to fight on modern battlefields.[26] From this portrayal of the future force, each of the services and the Department of Defense undertook efforts to transform their respective elements of the force. This meant a transformation across all elements of what became known as the DOTMLPF—doctrine, organizations, training, material, leadership, personnel, facilities. After the terrorist attacks of September 11, 2001, efforts to conduct a holistic transformation of the force were curtailed.

Specific DoD transformation activities that directly resulted from *JV 2010* have continued to advance and adapt to the evolving environment of the post–Cold War and post–9/11 world. Many of these efforts have led to new warfighting and support formations and processes. The DoD also has developed a centralized transformation program with numerous parts, including enterprise and business, information technology, and innovation transformation initiatives. Each service (i.e., army, navy, air force, marines) also has a transformation office. These transformation efforts seek to gain effectiveness and efficiency in the operational forces and support activities.

While transformation was clearly distinct from either S&T or R&D, each of the transformation initiatives envisioned by the Department of Defense and military services included elements that would have required these types of activities. For example, the army envisioned the fielding of a new type of network armored platform called the future combat system (FCS) that would have enhanced capabilities for command and control, precision fires at great distance, and enhanced protection through better situational awareness and be highly deployable. The development of such a platform depended on supporting scientific discovery and technology development. As described previously, this platform proved to be technologically infeasible and cost prohibitive.

APPENDIX B

TRL DEFINITIONS, DESCRIPTIONS, AND SUPPORTING INFORMATION FROM THE US DEPARTMENT OF DEFENSE[1]

[table begins on next page]

TRL	Definition	Description	Supporting Information
1	Basic principles observed and reported	Lowest level of technology readiness. Scientific research begins to be translated into applied research and development (R&D). Examples might include paper studies of a technology's basic properties.	Published research that identifies the principles that underlie this technology. References to who, where, when.
2	Technology concept and/or application formulated	Invention begins. Once basic principles are observed, practical applications can be invented. Applications are speculative, and there may be no proof or detailed analysis to support the assumptions. Examples are limited to analytic studies.	Publications or other references that outline the application being considered and that provide analysis to support the concept.
3	Analytical and experimental critical function and/or characteristic proof of concept	Active R&D is initiated. This includes analytical studies and laboratory studies to physically validate the analytical predictions of separate elements of the technology. Examples include components that are not yet integrated or representative.	Results of laboratory tests performed to measure parameters of interest and comparison to analytical predictions for critical subsystems. References to who, where, and when these tests and comparisons were performed.
4	Component and/or breadboard validation in laboratory environment	Basic technological components are integrated to establish that they will work together. This is relatively "low fidelity" compared with the eventual system. Examples include integration of "ad hoc" hardware in the laboratory.	System concepts that have been considered and results from testing laboratory-scale breadboard(s). Reference to who did this work and when. Provide an estimate of how breadboard hardware and test results differ from the expected system goals.

TRL	Definition	Description	Supporting Information
5	Component and/or breadboard validation in relevant environment	Fidelity of breadboard technology increases significantly. The basic technological components are integrated with reasonably realistic supporting elements so they can be tested in a simulated environment. Examples include "high-fidelity" laboratory integration of components.	Results from testing laboratory breadboard system are integrated with other supporting elements in a simulated operational environment. How does the "relevant environment" differ from the expected operational environment? How do the test results compare with expectations? What problems, if any, were encountered? Was the breadboard system refined to more nearly match the expected system goals?
6	System/subsystem model or prototype demonstration in a relevant environment	Representative model or prototype system, which is well beyond that of TRL 5, is tested in a relevant environment. Represents a major step up in a technology's demonstrated readiness. Examples include testing a prototype in a high-fidelity laboratory environment or in a simulated operational environment.	Results from a laboratory testing of a prototype system that is near the desired configuration in terms of performance, weight, and volume. How did the test environment differ from the operational environment? Who performed the tests? How did the test compare with expectations? What problems, if any, were encountered? What are/were the plans, options, or actions to resolve problems before moving to the next level?
7	System prototype demonstration in an operational environment	Prototype near or at planned operational system. Represents a major step up from TRL 6 by requiring demonstration of an actual system prototype in an operational environment (e.g., in an aircraft, in a vehicle, or in space).	Results from testing a prototype system in an operational environment. Who performed the tests? How did the test compare with expectations? What problems, if any, were encountered? What are/were the plans, options, or actions to resolve problems before moving to the next level?

TRL	Definition	Description	Supporting Information
8	Actual system completed and qualified through test and demonstration	Technology has been proven to work in its final form and under expected conditions. In almost all cases, this TRL represents the end of true system development. Examples include developmental test and evaluation (DT&E) of the system in its intended weapon system to determine if it meets design specification.	Results of testing the system in its final configuration under the expected range of environmental conditions in which it will be expected to operate. Assessment of whether it will meet its operational requirements. What problems, if any, were encountered? What are/were the plans, options, or actions to resolve problems before finalizing the design?
9	Actual system proven through successful mission operations	Actual application of the technology in its final form and under mission conditions, such as those encountered in operational test and evaluation (OT&E). Examples include using the system under operational mission conditions.	OT&E reports.

FRAMEWORK FOR ASSESSING TECHNOLOGY DEVELOPMENT AND AVAILABILITY

The Technology Assessment Methodology provides a structured approach to considering technology availability. For each framework area, assessment criteria serve to guide the analysis of a technology. Each area is composed of five components, an overarching assessment question, and levels that define the assessment scoring.

Using a decision support scoring methodology, the technologies can be assessed. A value from 1 to 5 can be selected signifying how a technology rates on each of the five components, and the results can be averaged in each of the five areas. Weighting can be employed to signify if an area or one of its components is of greater value in determining the availability of a technology. For example, in the case of nuclear material, the legal and regulatory impediments would greatly affect the ability to acquire fissile material, which would suggest the technology would not be readily available. The values for each of the five areas can be examined to determine the availability each contributes to the overall technology availability assessment.

Average scores of 5 signify a technology that is available and can be employed with little tacit knowledge or specialized equipment. Those technologies with average scores below 3 would indicate technologies that are not mature, could require specialized skills or equipment, and are not likely to be widely available. Average scores of 1 would indicate an immature technology that is not readily available for use.

Area	Science & Technology Maturity	Use Case, Demand & Market Forces	Resources Committed	Policy, Legal, Ethical & Regulatory Impediments	Technology Accessibility
Assessment	What is the level of maturity of the science and technology under consideration?	Is the technology likely to be an essential component or building block for advancement in other fields?	Will resource constraints serve as a barrier to development of the technology?	Are policy, legal, ethical, or regulatory barriers likely to serve as a barrier to the development of the technology?	Are the controls on the technologies likely to limit access to the wider population or be too technically sophisticated for widespread use?
	1. Elemental: theoretical or proof-of-concept 2. Initial use cases 3. Significant limitations 4. Modest limitations 5. Few/no limitations	1. Innovator use only 2. Early adopter use 3. Specialized uses only 4. Majority users 5. Broad use in variety of applications	Investment: 1. Little or none 2. Declining 3. Steady 4. Increasing 5. Exponential	Barriers to entry: 1. Absolute 2. Significant 3. Some 4. Few 5. None	Limitations: 1. Extreme 2. Significant 3. Some 4. Few 5. None
Components	Technology/ Knowledge/ Manufacturing Readiness Levels (TRL/KRL/MRL)	Perceived investment risk for technology development	Indirect investment (R&D, personnel, lab space)	Policy, legal, or ethical barriers	Cost to use
	Publications	Need or potential impact of innovation	Tech development cost (direct costs)	Export Controls	Regulatory
	Patents	Number and types of people/ groups using the technology	Total research dollars (denominator)	Technical leadership	Technical
	Miniaturization	Business case	Organization of investment	Existing regulations and ability to regulate	Who has access to tech
	Life-cycle improvements	Number and size of companies that own IP or are doing the R&D	Grants (government, philanthropic funding, etc.)	Comparative advantage	Democratization/ deskilling

OPERATIONAL VIEW DEFINITIONS FROM THE US DEPARTMENT OF DEFENSE

OV-1: High-Level Operational Concept Graphic—The high-level graphical/textual description of the operational concept.

OV-2: Operational Resource Flow Description—A description of the resource flows exchanged between operational activities.

OV-3: Operational Resource Flow Matrix—A description of the resources exchanged and the relevant attributes of the exchanges.

OV-4: Organizational Relationships Chart—The organizational context, role, or other relationships among organizations.

OV-5a: Operational Activity Decomposition Tree—The capabilities and activities (operational activities) organized in a hierarchal structure.

OV-5b: Operational Activity Model—The context of capabilities and activities (operational activities) and their relationships among activities, inputs, and outputs; Additional data can show cost, performers, or other pertinent information.

OV-6a: Operational Rules Model—One of three models used to describe activity (operational activity). It identifies business rules that constrain operations.

OV-6b: State Transition Description—One of three models used to describe operational activity (activity). It identifies business process (activity) responses to events (usually, very short activities).

OV-6c: Event-Trace Description—One of three models used to describe activity (operational activity). It traces actions in a scenario or sequence of events.

ELEMENTS OF THE DHS S&T PORTFOLIO REVIEW PROCESS (2011–2015)

Operational Focus

Operational Relevance: How well does the project's product(s) align with the customer's existing CONOP [concept of operation] or anticipated concept of use?

Customer Buy-in: Have the project objectives been developed through close consultation and continuing involvement with the people who will ultimately procure and transition the solution?

Innovation

Capability: To what extent does this project provide risk or threat reduction and/or improved fidelity, performance, etc.?

Efficiency: What level of savings can be achieved by this project with respect to the customer's operations?

Novel Approach: Does the project attempt to realize its objectives in a way that others have not previously considered or exploited?

Technical Feasibility: Is the project feasible given its most technically challenging aspect?

Partnerships

Resource Leverage: What level of commitment exists between the project team and the target component or end-user customer community? (Resource leverage may also be through interagency, international, academic, and/or industry relationships)

Foraging: Is the team taking advantage of everything that is known in the domain to facilitate a better, faster, cheaper solution?

Project Quality

Project Clarity: How well is the project described and laid out—is it clear what the team will do?

Cost Realism: Is the cost projection credible (given the scope of the project)? Is it clear how the cost projection was derived?

Timeline: When will the project achieve the efficiency or capability improvement as scored on the previous page?

Transition Likelihood: Is there a clear path and/or mechanism to enable transition or commercialization? Are there secondary issues related to customer readiness, concept of use, proponency, budgeting, affordability, regulatory or statutory realities, or business value?

CATEGORIES OF ACTIVITY FOR PROTECTING TECHNOLOGY

Category	Examples
International Laws & Treaties (including arms control)	• Nuclear Nonproliferation Treaty (NPT) • Biological Weapons Convention • Chemical Weapons Convention • UN Convention on Law of the Sea (UNCLOS) • Outer Space Treaty • UN Security Council Resolutions 1540 & 1874 • International sanctions
International Bodies	• Missile Technology Control Regime (MTCR) • Australia Group • Nuclear Suppliers Group (NSG) • World Health Organization (WHO) • INTERPOL
Coalitions of the Willing	• Proliferation Security Initiative (PSI) • Global Initiative to Combat Nuclear Terrorism (GICNT) • Wassenaar Arrangement • Blockades
US Initiatives	• Export Control & Border Security (EXBS) • Cooperative Threat Reduction (CTR) • Homeland Security & Counter Terrorism (CT) • Committee on Foreign Investment in the United States (CFIUS) • Customs and Border Protection's National Targeting Center

National Laws & Best Practices	• Laws, policies, and regulations • Intellectual property rights and patents • Standards and norms of behavior • Codes of ethics • Institutional R&D committees • Corporate and government proprietary information

ABBREVIATIONS AND INITIALISMS

3-D	three-dimensional
AC	alternating current
AEC	Atomic Energy Commission
AESA	Active Electronically Scanned Array
AFRL	Air Force Research Laboratory
AI	artificial intelligence
ARPA	Advanced Research Projects Agency
ARPA-E	Advanced Research Projects Agency-Energy
AT&T	American Telephone and Telegraph
AWACS	airborne warning and control system
BAA	Broad Agency Announcement
BARDA	Biomedical Advanced Research and Development Authority
BCE	Before Common Era
BP	British Petroleum
BSL	biological safety level
BWC	Biological Weapons Convention
CATIC	China National Aero-Technology Import and Export Corporation
CBP	Customs and Border Protection
CCL	Commerce Control List
CDMA	code-division multiple access
CE	Common Era
CFIUS	Committee on Foreign Investment in the United States
COE	centers of excellence
CPM	critical path method
CP-1	Chicago Pile 1
CRADA	cooperative R&D agreement

CRISPR	Clustered Regularly Interspaced Short Palindromic Repeats
CSI	Container Security Initiative
CTBT	Comprehensive Test Ban Treaty
C3	command, control, and communications
C-TPAT	Customs-Trade Partnership Against Terrorism
CTR	Cooperative Threat Reduction
CWC	Chemical Weapons Convention
DARPA	Defense Advanced Research Projects Agency
DC	direct current
DHS	Department of Homeland Security
DII	Defense Innovation Initiative
DIU	Defense Innovation Unit
DIUx	Defense Innovation Unit Experimental
DiY	do-it-yourself
DNA	deoxyribonucleic acid
DNI	Director of National Intelligence
DoD	Department of Defense
DoE	Department of Energy
DOTMLPF	doctrine, organization, training, material, leadership, personnel, facilities
DPRK	Democratic People's Republic of Korea
DURC	dual use research of concern
ENCODE	Encyclopedia of DNA Elements
ENIAC	Electronic Numerical Integrator and Computer
EVMS	earned value management system
FAAMG	Facebook, Amazon, Apple, Microsoft, and Google
FANG	Facebook, Amazon, Netflix, and Google
FAO	Food and Agriculture Organization of the United Nations
FCS	future combat system
FDA	Food and Drug Administration
FEMA	Federal Emergency Management Agency
FFRDC	federally funded research and development center
FMCT	Fissile Material Cutoff Treaty

GICNT	Global Initiative to Combat Nuclear Terrorism
GMO	genetically modified organism
GPS	Global Positioning System
HHS	Department of Health and Human Services
HPAI	highly pathogenic avian influenza
HPC	high-performance computing
HSARPA	Homeland Security Advanced Research Projects Agency
HSE	Homeland Security Enterprise
HSI	Homeland Security Investigations
IAEA	International Atomic Energy Agency
I&T	innovation and transformation
IARPA	Intelligence Advanced Research Projects Activity
IC	Intelligence Community
ICE	Immigration and Customs Enforcement
ICT	information and communication technologies
IEEE	Institute of Electrical and Electronics Engineers
INF	Intermediate Nuclear Force
IoT	internet of things
IP	intellectual property
IRAD	internal research and development
ISS	International Space Station
IT	information technology
ITAR	International Traffic in Arms Regulations
JCIDS	Joint Capabilities Integration and Development System
JROC	Joint Requirements Oversight Council
KRL	knowledge readiness level
MCS	maneuver control system
MEMS	micro-electromechanical systems
MRL	manufacturing readiness level
MTCR	Missile Technology Control Regime
NASA	National Aeronautics and Space Administration
NATO	North Atlantic Treaty Organization
NBACC	National Biodefense Analysis and Countermeasures Center

NIH	National Institutes of Health
NPT	Nuclear Nonproliferation Treaty
NSC	National Security Council
NSF	National Science Foundation
NSG	Nuclear Suppliers Group
NSTC	National Science and Technology Council
OIE	World Organization for Animal Health
OSRD	Office of Scientific Research and Development
OSTP	Office of Science and Technology Policy
OT	operational technology
OV	Operational View
PCAST	President's Council of Advisors on Science and Technology
PCR	polymerase chain reaction
PERT	Program Evaluation Review Technique
PPBES	Planning, Programming, Budgeting, and Execution System
PPBS	Planning, Programming, and Budgeting System
PSI	Proliferation Security Initiative
R&D	research and development
RFID	radio frequency identification
RMA	Revolution in Military Affairs
ROI	return on investment
S&T	science and technology
SBIR	Small Business Innovative Research
START	Strategic Arms Reduction Treaty
STEM	science, technology, engineering, and mathematics
STTR	Small Business Technology Transfer
TCP/IP	Transmission Control Protocol / Internet Protocol
TEPCO	Tokyo Electric Power Company
3-D	three-dimensional
TNT	trinitrotoluene
TRIZ	Theoria Resheneyva Isobretatelskehuh Zadach
TRL	technology readiness level

TSA	Transportation Security Administration
USML	US Munitions List
VCs	venture capital companies
WHO	World Health Organization
WMD	weapon of mass destruction
WTO	World Trade Organization

NOTES

CHAPTER 1. IN THE BEGINNING . . .

1. "Stone Tools," Smithsonian National Museum of Natural History, last updated March 5, 2019, http://www.humanorigins.si.edu/evidence/behavior/stone-tools.

2. Katherine Harmon, "Humans Feasting on Grains for at Least 100,000 Years," *Scientific American*, December 17, 2009, https://blogs.scientificamerican.com/observations/humans-feasting-on-grains-for-at-least-100000-years/.

3. Sindya N. Bhanoo, "Remnants of an Ancient Kitchen Are Found in China," *New York Times*, June 28, 2012, http://www.nytimes.com/2012/07/03/science/oldest-known-pottery-found-in-china.html.

4. Matthew Mason, "History of Agriculture," Environmental Science.org, https://www.environmentalscience.org/history-agriculture.

5. Laurie L. Dove, "5 Farming Techniques That Changed the World," How Stuff Works, May 30, 2014, https://science.howstuffworks.com/environmental/green-science/5-farming-technologies-changed-world.htm.

6. Ibid.

7. "'World's Oldest Calendar' Discovered in Scottish Field," BBC News, July 15, 2013, http://www.bbc.com/news/uk-scotland-north-east-orkney-shetland-23286928.

8. "Edward Jenner (1749–1823)," Brought to Life: Exploring the History of Medicine, http://broughttolife.sciencemuseum.org.uk/broughttolife/people/edwardjenner.

CHAPTER 2. PERSPECTIVES AND DEFINITIONS

1. "Major Periods in World History," Biography Online, https://www.biographyonline.net/different-periods-in-history/.

2. Elizabeth Palermo, "Who Invented the Printing Press?" Live Science, February 25, 2014, https://www.livescience.com/43639-who-invented-the-printing-press.html.

3. Chris Butler, "The Invention of the Printing Press and Its Effects," Flow of History, 2007, http://www.flowofhistory.com/units/west/11/FC74.

4. Rosalie E. Leposky, "A Brief History of Electricity," *Electrical Contractor*, January 2000, https://www.ecmag.com/section/your-business/brief-history-electricity.

5. "A Nuclear-Powered World," NPR, May 16, 2011, https://www.npr.org/2011/05/16/136288669/a-nuclear-powered-world.

6. "Nuclear Power in the World Today," World Nuclear Association, last updated February 2019, http://www.world-nuclear.org/information-library/current-and-future-generation/nuclear-power-in-the-world-today.aspx.

7. *Nuclear Posture Review* (Washington, DC: US Department of Defense, February 2018), https://media.defense.gov/2018/Feb/02/2001872886/-1/-1/1/2018-NUCLEAR-POSTURE-REVIEW-FINAL-REPORT.PDF.

8. *Nuclear Deference in the 21st Century: An Enduring Necessity* (Washington, DC: US Department of Defense, 2018), https://media.defense.gov/2018/Feb/02/2001872879/-1/-1/1/NUCLEAR-DETERRENCE.PDF.

9. Homeland Security Act of 2002, Pub. L. No. 107–296, 116 Stat. 2135 (November 25, 2002), https://www.dhs.gov/xlibrary/assets/hr_5005_enr.pdf.

10. John Butler-Adam, "The Weighty History and Meaning behind the Word 'Science,'" *Conversation*, October 1, 2015, http://theconversation.com/the-weighty-history-and-meaning-behind-the-word-science-48280.

11. "Our Definition of Science," Science Council, 2019, https://sciencecouncil.org/about-science/our-definition-of-science/; Ian Sample, "What Is This Thing We Call Science? Here's One Definition . . .," *Guardian*, March 3, 2009, https://www.theguardian.com/science/blog/2009/mar/03/science-definition-council-francis-bacon.

12. Appendix B provides definitions for the nine technology readiness levels employed by the US Department of Defense for this purpose.

13. Nick Skillicorn, "What Is Innovation? 15 Experts Share Their Innovation," Idea to Value, March 18, 2016, https://www.ideatovalue.com/inno/nickskillicorn/2016/03/innovation-15-experts-share-innovation-definition/.

14. *Business Dictionary*, s.v. "Innovation," http://www.businessdictionary.com/definition/innovation.html.

15. These themes come from an article: Daniel M. Gerstein, "The Military's Search for Innovation," RealClearDefense, August 13, 2018, https://www.realcleardefense.com/articles/2018/08/13/the_militarys_search_for_innovation_113718.html.

16. Paul K. Davis, "Military Transformation? Which Transformation, and What Lies Ahead?," in *The George W. Bush Defense Program: Policy, Strategy, and War*, ed. Stephen Cimbala (Washington, DC: Potomac Books, 2010), 11–41, https://www.rand.org/content/dam/rand/pubs/reprints/2010/RAND_RP1413.pdf.

17. Ibid.

18. All data presented in this section come from "Research and Development: US Trends and International Comparisons," chap. 4 of *Science & Engineering Indicators 2018* (Alexandria, VA: National Science Board, National Science Foundation, 2018), https://www.nsf.gov/statistics/2018/nsb20181/assets/1038/research-and-development-u-s-trends-and-international-comparisons.pdf, except as specifically noted.

19. "Overview of NCSES Surveys," National Science Foundation, last updated September 19, 2018, https://www.nsf.gov/statistics/srvyoverview/index.cfm. The National Patterns data derive from six major surveys of the organizations that perform or fund the bulk of US R&D:

Business R&D and Innovation Survey, Higher Education Research and Development Survey, Survey of Federal Funds for Research and Development, Federally Funded Research and Development Center R&D Survey, Survey of State Government Research and Development, and Survey of Research and Development Funding and Performance by Nonprofit Organizations.

20. "Research and Development," p. 88.

21. Ibid., p. 90.

22. *Guide to the President's Budget: Research & Development FY 2017* (Washington, DC: American Association for the Advancement of Science, September 20, 2016), https://www .aaas.org/sites/default/files/AAAS%20R&D%20Report%20FY17%20web.pdf.

CHAPTER 3. FOUNDATIONS OF TECHNOLOGY DEVELOPMENT

1. Clayton M. Christensen, "Exploring the Limits of the Technology S-Curve," *Production and Operations Management* 1, no. 4 (Fall 1992), https://pdfs.semanticscholar.org/ 31b5/f477879457c513aba2327c8225046c2a3d96.pdf.

2. *Encyclopedia Britannica*, s.v. "Gene Editing," https://www.britannica.com/science/ gene-editing. CRISPR stands for clustered regularly interspaced short palindromic repeats. More on CRISPR technology and associated Cas9 proteins will be discussed later.

3. Peter Jakab, "Leonardo da Vinci and Flight," Smithsonian National Air and Space Museum, August 22, 2013, https://airandspace.si.edu/stories/editorial/leonardo-da -vinci-and-flight.

4. Orville and Wilbur Wright, Flying Machine, US Patent 821,393, filed March 23, 1903, and issued May 22, 1906. Available online on the National Air and Space Museum website, https://airandspace.si.edu/exhibitions/wright-brothers/online/fly/1903/patenting.cfm.

5. It appears that Henry Ford never actually said this, but the phrase and associated sentiment remain attributed to him. Patrick Vlaskovits, "Henry Ford, Innovation, and That 'Faster Horse' Quote," *Harvard Business Review*, August 29, 2011, https://hbr.org/2011/08/ henry-ford-never-said-the-fast.

6. Erik Sass, "12 Technological Advancements of World War I," *Mental Floss*, April 30, 2017, http://mentalfloss.com/article/31882/12-technological-advancements-world-war-I.

7. Chris Finnamore and David Ludlow, "Top Inventions and Technical Innovations of World War 2," *Expert Reviews*, May 1, 2015, http://www.expertreviews.co.uk/ technology/7907/top-inventions-and-technical-innovations-of-world-war-2/page/0/4.

8. "Vannevar Bush," Internet Pioneers, https://www.ibiblio.org/pioneers/bush.html.

9. "Niels Bohr: Biographical," Nobel Prize, https://www.nobelprize.org/nobel_prizes/ physics/laureates/1922/bohr-bio.html, From *Nobel Lectures, Physics 1922–1941* (Amsterdam: Elsevier, 1965).

10. C. N. Trueman, "Louis Pasteur," History Learning Site, March 17, 2015, https:// www.historylearningsite.co.uk/a-history-of-medicine/louis-pasteur/.

11. Martin Enserink, "Free to Speak, Kawaoka Reveals Flu Details While Fouchier Stays Mum," *Science*, April 3, 2012, https://www.sciencemag.org/news/2012/04/ free-speak-kawaoka-reveals-flu-details-while-fouchier-stays-mum.

12. Ibid.

13. "Dual Use Research of Concern," Office of Science Policy, National Institutes of Health, https://osp.od.nih.gov/biotechnology/dual-use-research-of-concern/.

14. *United States Government Policy for Institutional Oversight of Life Sciences Dual Use Research of Concern* (Washington, DC: Public Health Emergency, US Department of Health and Human Services, September 2014), http://www.phe.gov/s3/dualuse/Documents/durc -policy.pdf.

15. "Biomedical Advanced Research and Development Authority," US Department of Health and Human Services, https://www.phe.gov/about/BARDA/Pages/default.aspx.

16. "Disruptive Technology," WhatIs.com, http://whatis.techtarget.com/definition/ disruptive-technology.

17. James Manyika et al., "Disruptive Technologies: Advances That Will Transform Life, Business, and the Global Economy," McKinsey Global Initiative, May 2013, https:// www.mckinsey.com/business-functions/digital-mckinsey/our-insights/disruptive-technologies.

18. Ibid.

19. Manyika, "Disruptive Technologies."

20. Daniel M. Gerstein, "Will a Robot Take My Job?," *National Interest Magazine*, April 18, 2018, http://nationalinterest.org/feature/will-robot-take-my-job-25444?page=show.

21. Inmarsats were not cellular telephones but rather satellite systems that were mobile and allowed for communications in austere environments. See "Our Satellites," Inmarsat, https://www.inmarsat.com/about-us/our-satellites/.

22. Tuan C. Nguyen, "The History of Smartphones," ThoughtCo., September 10, 2018, https://www.thoughtco.com/history-of-smartphones-4096585.

23. "Ericsson Mobility Report: 70 Percent of World's Population Using Smartphones by 2020," Ericsson Corporation, June 3, 2015, https://www.ericsson.com/en/press-releases/ 2015/6/ericsson-mobility-report-70-percent-of-worlds-population-using-smartphones -by-2020.

24. Bernd Lukasch, "From Lilienthal to the Wrights," Otto Lilienthal Museum, 2003, http://www.lilienthal-museum.de/olma/ewright.htm.

25. Remy Melina, "The Fallen Heroes of Human Spaceflight," *LiveScience*, April 1, 2011, https://www.livescience.com/33175-the-fallen-heroes-of-human-spaceflight.html.

26. "About Project 112 and Project SHAD," Public Health, US Department of Veterans Affairs, https://www.publichealth.va.gov/exposures/shad/basics.asp.

27. "US Public Health Service Syphilis Study at Tuskegee," Centers for Disease Control and Prevention, https://www.cdc.gov/tuskegee/timeline.htm.

28. "The Hindenburg Disaster," Airships.net, http://www.airships.net/hindenburg/ disaster/.

29. Vicki Bassett, "Causes and Effects of the Rapid Sinking of the Titanic," *Penn State Undergraduate Engineering Review*, November 1998, http://writing.engr.psu.edu/uer/bassett .html.

30. Johan Brown-Cohen, "Cascading Power Failures," Physics 240, Stanford University, October 23, 2010, http://large.stanford.edu/courses/2010/ph240/brown-cohen1/.

CHAPTER 4. THE MODERN AGE OF TECHNOLOGY

1. Matt Hourihan, "Science and Technology Funding under Obama: A Look Back," American Association for the Advancement of Science, January 18, 2017, https://www.aaas.org/news/science-and-technology-funding-under-obama-look-back.

2. Steve Blank, "The Endless Frontier: US Science and National Industrial Policy (Part 1)," *Berkeley Blog*, University of California, Berkeley, January 14, 2013, http://blogs.berkeley.edu/2013/01/14/the-endless-frontier-u-s-science-and-national-industrial-policy-part-1/.

3. Vannevar Bush, *Science, the Endless Frontier: A Report to the President* (Washington, DC: US Government Printing Office, July 1945), https://www.nsf.gov/od/lpa/nsf50/vbush1945.htm#letter.

4. Ibid.

5. Blank, "Endless Frontier."

6. Bush, *Science*.

7. "A Timeline of NSF History," National Science Foundation, https://www.nsf.gov/about/history/overview-50.jsp.

8. James G. Hershberg, *James B. Conant: Harvard to Hiroshima and the Making of the Nuclear Age* (New York: Knopf, 1993), pp. 305–9.

9. Dimitri Kusnezov, "The Department of Energy's National Laboratory Complex" (PowerPoint presentation, Commission to Review the Effectiveness of National Energy Laboratories, US Department of Energy, Washington, DC, July 18, 2014), https://energy.gov/sites/prod/files/2014/08/f18/July%2018%20Kusnezov%20FINAL.pdf .

10. Ibid.

11. "The Truman Doctrine," Office of the Historian, US Department of State, https://history.state.gov/milestones/1945-1952/truman-doctrine.

12. "*Sputnik*, 1957," Office of the Historian, US Department of State, https://history.state.gov/milestones/1953-1960/sputnik.

13. Paul Dickson, "Sputnik's Impact on America," *Nova* (blog post), PBS, November 6, 2007, http://www.pbs.org/wgbh/nova/space/sputnik-impact-on-america.html.

14. Ibid.

15. President Eisenhower also announced the Eisenhower Doctrine in January 1957, which made provisions for a country to request American economic assistance and/or aid from US military forces if it was being threatened by armed aggression from another state in order "to secure and protect the territorial integrity and political independence of such nations, requesting such aid against overt armed aggression from any nation controlled by international communism." See "The Eisenhower Doctrine, 1957," Office of the Historian, US State Department, https://history.state.gov/milestones/1953-1960/eisenhower-doctrine.

16. "Dwight D. Eisenhower and Science & Technology," Dwight D. Eisenhower Memorial Commission, November 2008, https://web.archive.org/web/20101027163454/http://eisenhowermemorial.org/onepage/IKE%20%26%20Science.Oct08.EN.FINAL%20%28v2%29.pdf (site discontinued and retrieved via Internet Archive Wayback Machine).

17. Ibid.

18. "About DARPA," Defense Advanced Research Projects Agency, https://www.darpa.mil/about-us/about-darpa.

19. Ibid.

20. "The Heilmeier Catechism," Defense Advanced Research Projects Agency, https://www.darpa.mil/work-with-us/heilmeier-catechism.

21. "Defense Laboratory Enterprise (DLE) Map," Defense Laboratory Enterprise, Office of the Assistant Secretary of Defense for Research and Engineering, https://www.acq.osd.mil/rd/laboratories/labs/map.html.

22. "Air Force Research Laboratory," US Air Force, December 15, 2014, http://www.af.mil/About-Us/Fact-Sheets/Display/Article/104463/air-force-research-laboratory/.

23. "ARPA-E Budget," Advanced Research Project Agency-Energy, US Department of Energy, https://arpa-e.energy.gov/?q=arpa-e-site-page/arpa-e-budget.

24. "ARPA-E Impact," Advanced Research Project Agency-Energy, US Department of Energy, https://arpa-e.energy.gov/?q=site-page/arpa-e-impact.

25. "Research Programs," IARPA, Office of the Director of National Intelligence, https://www.iarpa.gov/index.php/research-programs.

26. The HSE is a term that includes DHS; the operating components in the department; state, local, tribal, and territorial (SLTT) governments; law enforcement and first responders; and critical infrastructure sectors in the United States.

27. *DHS Science and Technology Directorate* (Washington, DC: Homeland Security Advanced Research Projects Agency, October 2016), https://www.dhs.gov/sites/default/files/publications/HSARPA%20Fact%20Sheet-508%20Oct16.pdf.

28. *FY 2018 Budget in Brief* (Washington, DC: US Department of Homeland Security, 2018), https://www.dhs.gov/sites/default/files/publications/DHS%20FY18%20BIB%20Final.pdf.

29. "Budget," National Institutes of Health, Department of Health and Human Services, https://www.nih.gov/about-nih/what-we-do/budget.

30. "About the National Science Foundation," National Science Foundation, https://www.nsf.gov/about/.

31. "About Funding," National Science Foundation, https://www.nsf.gov/funding/aboutfunding.jsp. The most recent NSF Strategic Plan was released in February 2018 and covers the period of 2018–2022: https://www.nsf.gov/pubs/2018/nsf18045/nsf18045.pdf?WT.mc_id=USNSF_124.

32. "Defense Innovation Marketplace," US Department of Defense, http://www.defenseinnovationmarketplace.mil/DII_Defense_Innovation_Initiative.html. The first offset strategy was in the 1950s; the Soviet Union fielded a much larger military force in Eastern Europe than NATO did. To offset this asymmetry, the United States fielded an arsenal of nuclear weapons, building a triad consisting of bombers, missiles, and submarines to deter any Soviet intention of invading Western Europe. The second offset occurred as a result of the development of precision-guided conventional weapons employed under the new operational concept of joint operations. The key drivers of this strategy were precision, stealth, and an operational concept that focused on achieving desired effects as opposed to simply destruction. The third offset seeks to improve the battle network's performance

by exploiting advances in artificial intelligence and autonomy, allowing the joint force to assemble and operate advanced joint collaborative human-machine battle networks of even greater power. The Department of Defense asked for $3.6 billion in the 2017 defense budget to begin research and development for the third offset capabilities we will need to meet the new security challenges. Larry Spencer, "US Air Force Key to Third Offset Strategy," *Defense News*, November 7, 2016, https://www.defensenews.com/opinion/commentary/2016/11/07/us-air-force-key-to-third-offset-strategy/.

33. "Defense Innovation Unit," US Department of Defense, https://diux.mil/team.

34. US Department of Defense, "MD5—A New Department of Defense National Security Technology Accelerator—Officially Launches with Disaster Relief Hackathon in New York City," news release no. NR-368-16, October 14, 2016. https://www.defense.gov/News/News-Releases/News-Release-View/Article/974626/md5-a-new-department-of-defense-national-security-technology-accelerator-offici/.

35. "Idea Lab," US Department of Health and Human Services, https://www.hhs.gov/idealab/about/.

36. "World's Top 10 Innovation Hubs," Business Facilities, March 6, 2017, https://businessfacilities.com/2017/03/worlds-top-10-innovation-hubs/.

37. *Technology Hubs* (Bonn, Germany: German Federal Ministry for Economic Cooperation and Development, 2013), http://10innovations.alumniportal.com/fileadmin/10innovations/dokumente/GIZ_10innovations_Technology-Hubs_Brochure.pdf.

38. *The Technology Reinvestment Project: Integrating Military and Civilian Industries* (Washington, DC: Congressional Budget Office, July 1993), https://www.cbo.gov/sites/default/files/103rd-congress-1993-1994/reports/93doc158.pdf.

39. Nicolas Friederici, "What Is a Tech Innovation Hub Anyway?" Oxford Internet Institute, University of Oxford, September 16, 2014, https://dig.oii.ox.ac.uk/2014/09/16/what-is-a-tech-innovation-hub-anyway/.

40. In the DHS, the operational organizations are referred to as the components. They include the Customs and Border Protection (CBP), Citizenship and Immigration Service (CIS), Federal Emergency Management Agency (FEMA), Immigration and Customs Enforcement (ICE), US Coast Guard (USCG), Transportation Security Administration (TSA), and the US Secret Service (USSS). Other major customers for support include the Domestic Nuclear Development Office (DNDO), the Office of Health Affairs (OHA), the Office of the Under Secretary for Management (USM), the Intelligence and Analysis Directorate (I&A), and the National Protection and Programs Directorate (NPPD).

41. Jon Gertner, *The Idea Factory: Bell Labs and the Great Age of the American Inventor* (New York: Penguin, 2012), p. 35.

42. "Transportation Security Laboratory," Science & Technology Directorate, US Department of Homeland Security, https://www.dhs.gov/science-and-technology/transportation-security-laboratory; "National Biodefense Analysis and Countermeasures Center," Science & Technology Directorate, US Department of Homeland Security, https://www.dhs.gov/science-and-technology/national-biodefense-analysis-and-countermeasures-center.

43. Some of the colleges and universities have been designated as UARCs, while others

have a less established relationship with the government and others conducting research and development.

44. Michael Lombardi (Boeing Corporation), in discussion with the author, March 2, 2018.

45. Boeing Corporation engineers, in discussion with the author, March 16, 2018.

46. Ibid.

47. Ibid.

48. Luis E. Romero, "Elon Musk's 4 Success Mantras from SpaceX and Tesla," Forbes, October 17, 2017, https://www.forbes.com/sites/luisromero/2017/10/17/elon-musks-4 -success-mantras-from-spacex-and-tesla/#1ceea365382f.

49. Samantha Masunaga, "SpaceX Nears Launch of Falcon Heavy, Facing a Changing Market for Heavy-Lift Rockets," *Los Angeles Times*, February 5, 2018, http://www.latimes .com/business/la-fi-spacex-falcon-heavy-20180205-story.html.

50. Ibid.

51. "Chapter 4: Research and Development: US Trends and International Comparisons," in *Science & Engineering Indicators 2018* (Alexandria, VA: National Science Board, National Science Foundation, 2018), https://www.nsf.gov/statistics/2018/nsb20181/assets/1038/ research-and-development-u-s-trends-and-international-comparisons.pdf.

CHAPTER 5. LEADING AT THE SPEED OF TECHNOLOGY

1. Einstein to Roosevelt, Long Island, August 2, 1939, atomicarchive.com, National Science Digital Archive, http://www.atomicarchive.com/Docs/Begin/Einstein.shtml.

2. "Manhattan Project Spotlight: Enrico Fermi," Atomic Heritage Foundation, October 28, 2015, https://www.atomicheritage.org/article/manhattan-project -spotlight-enrico-fermi.

3. "Nuclear Weapons Timeline," International Campaign to Abolish Nuclear Weapons, http://www.icanw.org/the-facts/the-nuclear-age/.

4. John F. Kennedy, "Excerpt from the 'Special Message to the Congress on Urgent National Needs,'" speech to Congress, May 25, 1961, transcript posted by NASA, May 24, 2004, https://www.nasa.gov/vision/space/features/jfk_speech_text.html.

5. John F. Kennedy, "Rice Stadium Moon Speech," speech at Rice University, September 12, 1962, transcript posted by NASA, https://er.jsc.nasa.gov/seh/ricetalk.htm.

6. *Joint Vision 2010* (Washington, DC: Office of the Chairman of the Joint Chiefs of Staff, 1995), http://webapp1.dlib.indiana.edu/virtual_disk_library/index.cgi/4240529/ FID378/pdfdocs/2010/Jv2010.pdf.

7. Dan Goure, "Defense Firms Are Ready to Invest but Pentagon Must Make Good on Acquisition Reforms," *National Interest* (blog), February 12, 2018, http://nationalinterest.org/ blog/the-buzz/defense-firms-are-ready-invest-pentagon-must-make-good-24467.

8. Ibid.

9. Wikipedia, s.v. "Defense Innovation Unit," last modified April 6, 2019, 21:22, https://en.wikipedia.org/wiki/Defense_Innovation_Unit.

10. Terri Moon Cronk, "Shanahan: New Army Command Is 'The Future,'" US Department of Defense, October 10, 2018, https://www.defense.gov/explore/story/Article/1658985/shanahan-new-army-command-is-the-future/.

11. "F-16 Block 70," Lockheed Martin, https://www.lockheedmartin.com/us/products/f16/F-16-Block-70.html.

12. When I served in DHS S&T as the Deputy Under Secretary under the leadership of Dr. Tara O'Toole, this concept became an inherent part of the research and development for the department.

CHAPTER 6. ASSESSING TECHNOLOGY

1. US Department of Defense, *The Military Critical Technologies List, Part II: Weapons of Mass Destruction Technologies* (Washington, DC: Office of the Under Secretary of Defense for Acquisition and Technology, 1998), p. II-3-4.

2. Carla Tardi, "Moore's Law," Investopedia, last modified March 24, 2019, https://www.investopedia.com/terms/m/mooreslaw.asp.

3. Committee on Forecasting Future Disruptive Technologies, "Chapter 2: Existing Technology Forecasting Methodologies," in *Persistent Forecasting of Disruptive Technologies*, pp. 17–32 (Washington, DC: National Academies Press, 2010), https://www.nap.edu/read/12557/chapter/4.

4. "E-3 AWACS (Sentry) Airborne Warning and Control System," Air Force Technology, https://www.airforce-technology.com/projects/e3awacs/.

5. The four forecasting methods and their descriptions are from Committee on Forecasting, "Chapter 2."

6. "TRIZ: Theory of Inventive Problem Solving" (PowerPoint presentation, Department of Chemical Engineering, University of Michigan, Ann Arbor, 2011), http://umich.edu/~scps/html/07chap/html/powerpointpicstriz/Chapter%207%20TRIZ.pdf.

7. Committee on Forecasting, "Chapter 2," pp. 23–24.

8. This Technology Assessment Methodology was developed as part of a RAND study led by Daniel M. Gerstein examining how biotechnology improvements could contribute to the capabilities of a potential bioterrorist. The work was also based on the framework developed by Gerstein for use in a course taught at American University in Washington, DC, from 2015–2018 titled "Technology Policy and National Security."

9. "General Information concerning Patents," US Patent and Trademark Office, October 2015, https://www.uspto.gov/patents-getting-started/general-information-concerning-patents; "Patent Eligibility Requirements FAQ," Small Business Law, http://smallbusiness.findlaw.com/intellectual-property/patent-eligibility-requirements-faq.html.

CHAPTER 7. TECHNOLOGY DEVELOPMENT METHODS

1. Earlier in the chapter, the PPBES was discussed. The PPBS discussed here is the earlier version of the PPBES. The execution or "E" was added in the 1980s to reflect the growing desire to track the execution of the planning, programming, and budgeting that had been completed.

2. P. A. DonVito, *The Essentials of a Planning-Programming-Budgeting System* (Santa Monica, CA: RAND Corporation, 1969), https://www.rand.org/pubs/papers/P4124.html.

3. Note that the US Marine Corps is managed within the Department of the Navy.

4. DonVito, *Essentials*, pp. 1–2.

5. "10 US Code § 181—Joint Requirements Oversight Council," Legal Information Institute, Cornell University, https://www.law.cornell.edu/uscode/text/10/181.

6. Combatant commander refers to the senior warfighting commands within the military. There are ten combatant commands, six of which have geographic areas of responsibility (AOR). The regional commands include Central Command, Africa Command, European Command, Pacific Command, Northern Command, and Southern Command. The functional COCOMs are Special Operations Command, Transportation Command, and Strategic Command and Cyber Command. http://www.dodlive.mil/2016/08/17/dods-9-combatant-commands-what-they-are-what-they-do/.

7. The term technology development really implies R&D programs. Whether conducting technology or capability development, the R&D enterprise with the DoD is the system that ensures focused and connected programs are being conducted.

8. Thomas Hornigold, "How Fukushima Changed Japanese Robotics and Woke Up the Industry," SingularityHub, April 25, 2018, https://singularityhub.com/2018/04/25/how-fukushima-changed-japanese-robotics-and-woke-up-the-industry/#sm.00000409xfx64xfnlw6env0vgat3f.

9. "DARPA Robotics Challenge (DRC) (Archived)," Defense Advanced Research Projects Agency, https://www.darpa.mil/program/darpa-robotics-challenge.

10. "Small Business Innovative Research," US Small Business Administration, https://www.sbir.gov/about/about-sbir.

11. "About Small Business Innovative Research," US Small Business Administration, https://www.sbir.gov/about.

12. "About Small Business Technology Transfer (STTR)," US Small Business Administration, https://www.sbir.gov/about/about-sttr.

13. "Broad Area Announcements (BAA)," US Agency for International Development, https://www.usaid.gov/partnership-opportunities/respond-solicitation/broad-agency-announcements.

14. "Frequently Asked Questions: CRADAs," Technology Transfer Office, US Naval Research Laboratory, https://www.nrl.navy.mil/techtransfer/FAQs/CRADA.

15. "The Ultimate Guide to the Critical Path Method," Smartsheet, https://www.smartsheet.com/critical-path-method.

16. "What is EVM?," College of Performance Management, http://www.mycpm.org/what-is-evm/.

17. "DoDAF [Architectural Frameworks] Viewpoints and Models," Chief Information Officer, US Department of Defense, http://dodcio.defense.gov/Library/DoD-Architecture-Framework/dodaf20_ov1/.

18. Ibid.

19. The portfolio review process in this case was developed by Navigant. Other methodologies are certainly in use within the community. The use of such a tool allows for assessing the health of the portfolio both in terms of the individual projects and across the entire R&D portfolio.

CHAPTER 8. MANAGING TECHNOLOGY IN AN UNMANAGEABLE WORLD

1. A primary source for this introduction was Keith, "The History of Social Media: Social Networking Evolution!," History Cooperative, February 14, 2019, http://historycooperative.org/the-history-of-social-media/.

2. Hamza Shaban, Craig Timberg, and Elizabeth Dwoskin, "Facebook, Google and Twitter Testified on Capitol Hill. Here's What They Said," *Washington Post*, October 31, 2017, https://www.washingtonpost.com/news/the-switch/wp/2017/10/31/facebook-google-and-twitter-are-set-to-testify-on-capitol-hill-heres-what-to-expect/?utm_term=.c100637ced4e.

3. Matt Apuzzo and Sharon LaFraniere, "13 Russians Indicted as Mueller Reveals Effort to Aid Trump Campaign Image," *New York Times*, February 16, 2018, https://www.nytimes.com/2018/02/16/us/politics/russians-indicted-mueller-election-interference.html.

4. "Where Are GMOs Grown and Banned?," Genetic Literacy Project, https://gmo.geneticliteracyproject.org/FAQ/where-are-gmos-grown-and-banned/.

5. "18 US Code § 2332a—Use of weapons of mass destruction," Legal Information Institute, Cornell University, https://www.law.cornell.edu/uscode/text/18/2332a.

6. Robert H. Shmerling, "First, Do No Harm," Harvard Medical School, October 13, 2015, https://www.health.harvard.edu/blog/first-do-no-harm-201510138421.

7. "IEEE Code of Ethics," IEEE, https://www.ieee.org/about/corporate/governance/p7-8.html.

8. Sharon Florentine, "Should Software Developers Have a Code of Ethics?" *CIO*, March 30, 2018, https://www.cio.com/article/3156565/developer/should-software-developers-have-a-code-of-ethics.html.

9. "Intellectual Property Rights," OECD, January 3, 2002, https://stats.oecd.org/glossary/detail.asp?ID=3236.

10. "About WTO," World Trade Organization, https://www.wto.org/english/thewto_e/thewto_e.htm.

11. "Why Can't the US and China Agree on IP Rights?," Thomson Reuters, August 30, 2017, https://blogs.thomsonreuters.com/answerson/cant-u-s-china-agree-ip-rights/.

12. "US Charges Five Chinese Military Hackers for Cyber Espionage against US Corporations and a Labor Organization for Commercial Advantage," US Department of Justice, Office of Public Affairs, May 19, 2014, https://www.justice.gov/opa/pr/us-charges-five-chinese-military-hackers-cyber-espionage-against-us-corporations-and-labor.

13. Thomson Reuters, "IP Rights."

14. US Department of State, http://pmddtc.state.gov/ECR/index.html (page removed as of April 16, 2019). The twenty-one USML categories: Category I—Firearms, Close Assault Weapons/ Combat Shotguns; Category II—Guns and Armament; Category III—Ammunition/Ordnance; Category IV—Launch Vehicles, Guided Missiles, Ballistic Missiles, Rockets, Torpedoes, Bombs and Mines; Category V—Explosives and Energetic Materials, Propellants, Incendiary Agents and Their Constituents; Category VI—Vessels of War/Special Naval Equipment; Category VII—Tanks and Military Vehicles; Category VII—Aircraft and Associated Equipment; Category VIII— Military Training Equipment and Training; Category X—Protective Personnel; Equipment/ Shelters; Category XI—Military Electronics; Category XII—Fire Control, Range Finder, Optical and Guidance and Control Equipment; Category XIII—Auxiliary Military Equipment; Category XIV—Toxicological Agents, Including Chemical Agents, Biological Agents, and Associated Equipment; Category XV—Spacecraft Systems/Associated Equipment; Category XVI—Nuclear Weapons, Design and Testing Related Items; Category XVII—Classified Articles, Technical Data, and Defense Services Not Otherwise Enumerated; Category XVIII—Directed Energy Weapons; Category XIX [Reserved]; Category XX—Submersible Vessels, Oceanographic and Associated Equipment; Category XXI—Miscellaneous Articles.

15. US Department of State, http://pmddtc.state.gov/regulations_laws/documents/ official_itar/ITAR_Part_121.pdf (page removed as of April 16, 2019).

16. The ten categories: 0—Nuclear; 1—Materials, Chemicals, Microorganisms and Toxins; 2—Materials Processing; 3—Electronics; 4—Computers; 5—Telecommunications and Information Security; 6—Lasers and Sensors; 7—Navigation and Avionics; 8—Marine; and 9—Propulsion Systems, Space Vehicles and Related Equipment.

17. The five groups: A—Equipment, Assemblies and Components; B—Test, Inspection, and Production Equipment; C—Materials; D—Software; E—Technology.

18. Paul Koscak, "Working Together: Catching Smugglers, Terrorists, and Lawbreakers Works Better Through Partnership," US Customs and Border Protection, https://www.cbp .gov/frontline/cbp-national-targeting-center.

19. "Border Security," US Customs and Border Protection, https://www.cbp.gov/ border-security.

20. "Overview: Export Enforcement Coordination Center," US Immigration and Customs Enforcement, https://www.ice.gov/eecc.

21. "About: Office of Foreign Assets Control (OFAC)," US Department of the Treasury, https://www.treasury.gov/about/organizational-structure/offices/Pages/Office-of-Foreign -Assets-Control.aspx.

22. "US Export Controls," US Department of State, https://www.state.gov/ strategictrade/resources/c43182.htm.

23. Brenna Gautam, "Fact Sheet: The Nunn-Lugar Cooperative Threat Reduction Program," Center for Arms Control and Non-Proliferation, June 1, 2014, https:// armscontrolcenter.org/fact-sheet-the-nunn-lugar-cooperative-threat-reduction-program/.

24. James K. Jackson, *The Committee on Foreign Investment in the United States (CFIUS)* (Washington, DC: Congressional Research Service, July 3, 2018), https://fas.org/sgp/crs/ natsec/RL33388.pdf.

25. "Missile Technology Control Regime," NTI, April 1987 (last updated January 8, 2018), http://www.nti.org/learn/treaties-and-regimes/missile-technology -control-regime-mtcr/.

26. Nuclear Suppliers Group, http://www.nuclearsuppliersgroup.org/en/.

27. Australia Group, http://www.australiagroup.net/en/.

28. Wassenaar Arrangement, http://www.wassenaar.org/.

29. "Proliferation Security Initiative," US Department of State, https://www.state.gov/t/ isn/c10390.htm.

30. Global Initiative to Combat Nuclear Terrorism, http://www.gicnt.org/.

31. Information about the WHO, FAO, and OIE can be found at http://www.who.int/ en/, http://www.fao.org/animal-production/en/, and http://www.oie.int/, respectively.

32. "In just war theory, the decision to go to war should be governed by criteria including just cause, comparative justice, competent authority, right intention, probability of success, last resort, and proportionality. For actions during war, the following criteria should be considered in limiting actions: distinction, proportionality, and military necessity." From Wikipedia, s.v. "Just War Theory," last modified March 26, 2019, 00:45, https:// en.wikipedia.org/wiki/Just_war_theory.

33. "Convention (II) with Respect to the Laws and Customs of War on Land and Its Annex: Regulations concerning the Laws and Customs of War on Land. The Hague, 29 July 1899," Organization for the Prohibition of Chemical Weapons, http://www.opbw.org/int _inst/sec_docs/1899HC-TEXT.pdf.

34. "Biological Weapons Convention," United Nations, http://disarmament.un.org/ treaties/t/bwc/text.

35. "Treaties and Agreements," Arms Control Association, https://www.armscontrol .org/treaties.

36. "Kyoto Protocol—Targets for the First Commitment Period," United Nations Climate Change, http://unfccc.int/kyoto_protocol/items/2830.php; International Criminal Court, https://www.icc-cpi.int/.

37. Angus Maddison, "The West and the Rest in the World Economy: 1000–2030," ResearchGate, 2008, https://www.researchgate.net/publication/237678324_The_West _and_the_Rest_in_the_World_Economy_1000-2030_Maddisonian_and_Malthusian _interpretations; G. John Ikenberry, "Don't Panic: How Secure Is Globalization's Future?," *Foreign Affairs*, May/June 2000, https://www.foreignaffairs.com/reviews/review -essay/2000-05-01/dont-panic-how-secure-globalizations-future; both sources quoted in World Trade Organization, "Trends in International Trade," in *World Trade Report 2013: Factors Shaping the Future of World Trade* (Geneva: World Trade Organization, 2013), https:// www.wto.org/english/res_e/booksp_e/wtr13-2b_e.pdf.

38. Esteban Ortiz-Ospina, Diana Beltekian, and Max Roser, "Trade and Globalization," Our World in Data, 2018, https://ourworldindata.org/international-trade.

39. World Trade Organization, "Trends in International Trade," p. 71.

40. Ortiz-Ospina, Beltekian, and Roser, "Trade and Globalization."

41. Ibid.

42. Frank Cain, "Computers and the Cold War: United States Restrictions on the

Export of Computers to the Soviet Union and Communist China," *Journal of Contemporary History* 40, no. 1 (January 2005): 132, http://journals.sagepub.com/doi/pdf/10.1177/0022009405049270.

43. Ibid.

44. Ibid., 136–37.

45. Steve Greenhouse, "US and Allies Move to Ease Cold War Limits on Exports," *New York Times*, May 25, 1991, http://www.nytimes.com/1991/05/25/business/us-and-allies-move-to-ease-cold-war-limits-on-exports.html?pagewanted=all#h[CwrRac,2].

46. Markus Schiller and Robert H. Schmucker, "Flashback to the Past: North Korea's 'New' Extended-Range Scud," *38 North*, November 8, 2016, https://www.38north.org/wp-content/uploads/2016/11/Scud-ER-110816_Schiller_Schmucker.pdf.

47. Earlier, Israel had approached the United States about assisting in developing a nuclear program for peaceful purposes, which was refused unless nuclear safeguards were in place. Israel did not want to have safeguards on their nuclear material, as that would include inspections and reporting that undoubtedly would have telegraphed the nuclear weapons program.

48. Christopher Dickey, "How Shimon Peres Outwitted the US to Bring Nukes to Israel," Daily Beast, September 28, 2016, https://www.thedailybeast.com/how-shimon-peres-outwitted-the-us-to-bring-nukes-to-israel.

49. Volha Charnysh, "India's Nuclear Program," Nuclear Age Peace Foundation, September 3, 2009, http://nuclearfiles.org/menu/key-issues/nuclear-weapons/issues/proliferation/india/charnysh_india_analysis.pdf.

50. Leonard Weiss, "What Do Past Nonproliferation Failures Say about the Iran Nuclear Agreement?," *Bulletin of the Atomic Scientists*, September 1, 2015, https://thebulletin.org/what-do-past-nonproliferation-failures-say-about-iran-nuclear-agreement8706; Akhilesh Pillalamarri, "India's Nuclear-Weapons Program: 5 Things You Need to Know," *National Interest*, April 22, 2015, http://nationalinterest.org/feature/indias-nuclear-weapons-program-5-things-you-need-know-12697.

51. Catherine Collins and Douglas Frantz, "The Long Shadow of A. Q. Khan," *Foreign Affairs*, January 31, 2018, https://www.foreignaffairs.com/articles/north-korea/2018-01-31/long-shadow-aq-khan.

52. Ibid.

53. "North Korea's Latest Nuclear Test Yield Estimated at 250 Kilotons: US Monitor," *Straits Times*, September 13, 2017, http://www.straitstimes.com/asia/east-asia/north-koreas-latest-nuclear-test-yield-estimated-at-250-kilotons-us-monitor.

54. Daniel M. Gerstein, "How Did North Korea Get This Far?" *US News & World Report*, August 10, 2017, https://www.usnews.com/opinion/world-report/articles/2017-08-10/how-did-north-korea-get-its-nuclear-capabilities-so-far-so-fast.

55. "Report on North Korea's Activities," United Nations Security Council, February 27, 2017, http://www.un.org/ga/search/view_doc.asp?symbol=S/2017/150.

56. Dennis C. Blair and Keith Alexander, "China's Intellectual Property Theft Must Stop," *New York Times*, August 15, 2017, https://www.nytimes.com/2017/08/15/opinion/china-us-intellectual-property-trump.html.

57. Evan Perez, "US Charges 3 Chinese Nationals with Hacking, Stealing Intellectual Property from Companies," CNN, November 27, 2017, https://www.cnn.com/2017/11/27/politics/china-hacking-case/index.html.

58. The term applies especially to systems that function above a teraflop, or 10^{12} floating point operations per second.

59. Jack Corrigan, "17-Year-Old Hacks US Air Force for the Biggest Bug Bounty," Defense One, August 11, 2017, http://www.defenseone.com/technology/2017/08/17-year-old-hacks-air-force-biggest-bug-bounty/140165/.

CHAPTER 9. THINKING ORTHOGONALLY

1. "Members of Apollo 13 Team Reflect on 'NASA's Finest Hour,'" NASA, April 17, 2015, https://www.nasa.gov/content/members-of-apollo-13-team-reflect-on-nasas-finest hour.

2. Wikipedia, s.v. "Packet Switching," last modified April 15, 2019, 16:13, https://en.wikipedia.org/wiki/Packet_switching.

3. Margaret Rouse, "Definition: ARPANET," TechTarget, March 2017, http://searchnetworking.techtarget.com/definition/ARPANET.

4. Margaret Rouse, "Definition: Identity Management (ID Management)," TechTarget, November 2017, http://searchsecurity.techtarget.com/definition/identity-management -ID-management.

5. Rachel Cooper, "Washington DC Metropolitan Area Profile and Demographics," Tripsavvy, September 1, 2018, https://www.tripsavvy.com/washington-dc -area-profile-and-demographics-1038691.

6. "Coastal Flood Risks: Achieving Resilience Together," Federal Emergency Management Agency, https://www.fema.gov/coastal-flood-risks-achieving-resilience-together.

7. Adrian Marshall, "How Not to Screw Up Spending $1 Trillion on US Infrastructure," *Wired*, January 1, 2017, https://www.wired.com/2017/01/not-screw -spending-1-trillion-us-infrastructure/.

8. Ibid.

9. Ibid.

10. *Trends in Smart City Development* (Washington, DC: National League of Cities, 2016), https://www.nlc.org/sites/default/files/2017-01/Trends%20in%20Smart%20City%20 Development.pdf.

11. Teodora Zareva, "Saudi Arabia Is Building a $500-Billion New Territory Based on Tech and Liberal Values," Big Think, October 30, 2017, http://bigthink.com/design-for-good/saudi-arabia-is-building-a-utopian-city-to-herald-the-future-of-human-civillization.

12. National League of Cities, *Smart City Development*.

13. "NIH Partners with 11 Leading Biopharmaceutical Companies to Accelerate the Development of New Cancer Immunotherapy Strategies for More Patients," National Institutes of Health, October 12, 2017, https://www.nih.gov/news-events/news-releases/nih-partners-11-leading-biopharmaceutical-companies-accelerate-development-new

-cancer-immunotherapy-strategies-more-patients#overlay-context=institutes-nih/
nih-office-director.

14. W. H. Gerstenmaier, "Human Exploration & Operations Update on Mission
Planning for NAC," NASA, November 30, 2016, https://www.nasa.gov/sites/default/files/
atoms/files/human_exploration_operations_update_tagged.pdf.

CHAPTER 10. TECHNOLOGY'S EFFECT ON SOCIETY

1. Richard Conniff, "What the Luddites Really Fought Against," *Smithsonian*,
March 2011, https://www.smithsonianmag.com/history/what-the-luddites
-really-fought-against-264412/.

2. Ibid.

3. L. Patrick Hughes, "Agricultural Problems and Gilded Age Politics," Austin
Community College, http://www.austincc.edu/lpatrick/his1302/agrarian.html.

4. Derek Thompson, "The 100-Year March of Technology in 1 Graph," *Atlantic*,
April 7, 2012, https://www.theatlantic.com/technology/archive/2012/04/
the-100-year-march-of-technology-in-1-graph/255573/.

5. Steve Dennis, "E-Commerce May Be Only 10% of Retail, but That
Doesn't Tell the Whole Story," Forbes, April 9, 2018, https://www.forbes.com/sites/
stevendennis/2018/04/09/e-commerce-fake-news-the-only-10-fallacy/#26c12739b4fc.

6. "The 10 Biggest Dangers Posed by Future Technology," *CS Globe*, May 3, 2015,
http://csglobe.com/the-10-biggest-dangers-posed-by-future-technology/.

7. Ibid.

8. Milen Raychev, "The Dangers of Technology and Social Media," I Heart
Intelligence, October 4, 2018, http://iheartintelligence.com/2018/01/22/dangers
-technology-social-media/.

9. Peter Townsend, "How Dangerous is Technology?," *OUPBlog*, February 13, 2017,
https://blog.oup.com/2017/02/how-dangerous-is-technology/.

10. Tim Collins, "Ex-Google and Facebook Employees Launch a Campaign to Warn
about the Addictive Dangers of the Technologies They Helped Build," *Daily Mail*, February
5, 2018, http://www.dailymail.co.uk/sciencetech/article-5352867/Campaign-launched-warn
-dangers-technology.html.

11. "Darwin's Theory of Evolution," AllAboutScience.org, https://www.darwins-theory
-of-evolution.com/.

12. Tim Chao, Tuan Pham, and Mikhail Seregine, "The Dangers of Technological
Progress," Stanford University, https://cs.stanford.edu/people/eroberts/cs201/projects/
1999-00/technology-dangers/issues.html.

13. "NSF Strategic Plan for Fiscal Years (FY) 2018–2022," National Science
Foundation, https://www.nsf.gov/pubs/2018/nsf18045/nsf18045.pdf?WT.mc_id
=USNSF_124.

14. Rob Carlson, "Time for New DNA Synthesis and Sequencing Cost Curves,"

synbiobeta, February 17, 2014, https://synbiobeta.com/time-new-dna-synthesis
-sequencing-cost-curves-rob-carlson/.

15. While James Watson and Francis Crick have received much of the credit
for discovering DNA, this notoriety is misplaced. Their work was based on efforts by
several other scientists. Notably, Swiss chemist Friedrich Miescher in the late 1860s
discovered the nucleic acid compounds within cells. See Leslie A. Pray, "Discovery
of DNA Structure and Function: What Did Watson and Crick Actually Discover?,"
Nature Education 1, no. 1 (2008): 100, https://www.nature.com/scitable/topicpage/
discovery-of-dna-structure-and-function-watson-397.

16. Lydia Ramsey, "A Revolutionary Gene-Editing Technology Is on Track to Be a $10
Billion Market by 2025," *Business Insider*, November 2, 2017, https://www.businessinsider
.com/crispr-set-to-be-a-10-billion-market-by-2025-citi-2017-11.

17. "The History of PCR," Smithsonian Institution Archive, http://siarchives.si.edu/
research/videohistory_catalog9577.html.

18. David Whitehouse, "First Synthetic Virus Created," *BBC News*, July 11, 2002,
http://news.bbc.co.uk/2/hi/science/nature/2122619.stm.

19. Haley Bennett, "The First Synthetic Cell," Chemistry World, May 20, 2010,
https://www.chemistryworld.com/news/the-first-synthetic-cell/3003950.article.

20. "What Is the Encyclopedia of DNA Elements (ENCODE) Project?," US National
Library of Medicine, https://ghr.nlm.nih.gov/primer/genomicresearch/encode.

21. Amelia Gunnarsson, "Nuclear-Mitochondrial Mismatch: Divergent Evolution and
the Safety of Mitochondrial Replacement Therapy," *BMC Series Blog*, March 2, 2017, https://
blogs.biomedcentral.com/bmcseriesblog/2017/03/02/nuclear-mitochondrial-mismatch
-divergent-evolution-safety-mitochondrial-replacement-therapy/.

22. Sara Reardon, "Genetic Details of Controversial 'Three-Parent Baby' Revealed,"
Nature, April 3, 2017, https://www.nature.com/news/genetic-details-of
-controversial-three-parent-baby-revealed-1.21761.

23. Kathi E. Hanna, "Genetic Enhancement," National Human Genome Research
Institute, https://www.genome.gov/10004767/genetic-enhancement/.

24. Lorraine Chow, "16 European Nations Vote against GMO Crops," EcoWatch, March
27, 2017, https://www.ecowatch.com/eu-nations-vote-against-gmo-crops-2332080511.html.

25. A. S. Bawa and K. R. Anilakumar, "Genetically Modified Foods: Safety, Risks and
Public Concerns—A Review," *Journal of Food Science Technology* 50, no. 6 (December 2013):
1035–46, https://www.ncbi.nlm.nih.gov/pmc/articles/PMC3791249/.

26. Daniela Rus, "The Robots Are Coming," *Foreign Affairs*, July/August 2015.

27. *English Oxford Living Dictionary*, s.v. "Artificial Intelligence," https://en.oxford
dictionaries.com/definition/artificial_intelligence; Bernard Marr, "The Key Definitions of
Artificial Intelligence (AI) That Explain Its Importance," Forbes, February 14, 2018, https://
www.forbes.com/sites/bernardmarr/2018/02/14/the-key-definitions-of
-artificial-intelligence-ai-that-explain-its-importance/2/#4b4c11495284.

28. Dawn Chan, "The AI That Has Nothing to Learn from Humans," *Atlantic*,
October 20, 2017, https://www.theatlantic.com/technology/archive/2017/10/
alphago-zero-the-ai-that-taught-itself-go/543450/.

29. Ibid.; David Silver et al., "Mastering the Game of Go without Human Knowledge," *Nature*, October 19, 2017, https://www.nature.com/articles/nature24270.

30. Chan, "Nothing to Learn."

31. Paul Wiseman, "Why Robots, Not Trade, Are behind So Many Factory Job Losses," AP News, November 2, 2016, https://apnews.com/265cd8fb02fb44a69cf0eaa2063e11d9.

32. Scott Santens, "Self-Driving Trucks Are Going to Hit Us Like a Human-Driven Truck," Medium, May 14, 2015, https://medium.com/basic-income/self-driving-trucks-are-going-to-hit-us-like-a-human-driven-truck-b8507d9c5961.

33. These figures were current as of August 2017. See "Labor Force Statistics from the Current Population Survey," Bureau of Labor Statistics, US Department of Labor, https://data.bls.gov/timeseries/LNS12000000.

34. David R. Baker and Carolyn Said, "How the Bay Area Took Over the Self-Driving Car Business," *San Francisco Chronicle*, July 14, 2017, updated July 15, 2017, http://www.sfchronicle.com/business/article/How-the-Bay-Area-took-over-the-self-driving-car-11278238.php.

35. Ibid.

36. "Harnessing Automation for a Future That Works," McKinsey Global Institute, January 2017, http://www.mckinsey.com/global-themes/digital-disruption/harnessing-automation-for-a-future-that-works.

37. Defense Science Board, *Summer Study on Autonomy* (Washington, DC: US Department of Defense, 2016), https://www.hsdl.org/?view&did=794641.

38. Barry M. Leiner et al., "Brief History of the Internet," Internet Society, 1997, http://www.internetsociety.org/internet/what-internet/history-internet/brief-history-internet.

39. "World Stats," Internet World Stats, June 30, 2018, https://www.internetworldstats.com/stats.htm.

40. Margaret Rouse, "Definition: Internet of Things (IoT)," TechTarget, March 2019, http://internetofthingsagenda.techtarget.com/definition/Internet-of-Things-IoT.

41. *Strategic Principles for Securing the Internet of Things (IoT) (Version 1.0)* (Washington, DC: US Department of Homeland Security, November 15, 2016), https://www.dhs.gov/sites/default/files/publications/Strategic_Principles_for_Securing_the_Internet_of_Things-2016-1115-FINAL.pdf.

CHAPTER 11. THE TECHNOLOGIST'S RESPONSIBILITY TO SOCIETY

1. "RIP Kalashnikov: 20 Facts You May Not Have Known about AK-47 and Its Creator," *Russia Today*, December 23, 2013, https://www.rt.com/news/kalashnikov-rifle-ak47-facts-691/.

2. Ibid.

3. "Deepwater Horizon Drills World's Deepest Oil & Gas Well," Transocean, April 26, 2010, https://web.archive.org/web/20100426171257/http:/www.deepwater.com/fw/main/IDeepwater-Horizon-i-Drills-Worlds-Deepest-Oil-and-Gas-Well-419C151.html (page discontinued and retrieved via Internet Archive Wayback Machine).

4. "BP's Reckless Conduct Caused Deepwater Horizon Oil Spill, Judge Rules," *Guardian*, September 4, 2014, https://www.theguardian.com/environment/2014/sep/04/bp-reckless-conduct-oil-spill-judge-rules.

5. Justin Mullins, "Eight Failures That Caused the Gulf Oil Spill," *New Scientist*, September 8, 2010, https://www.newscientist.com/article/dn19425-the-eight-failures-that-caused-the-gulf-oil-spill/.

6. The term *large therapeutic window* denotes the range of medical countermeasure dosages that can treat disease effectively without having toxic effects.

7. "Backgrounder on the Three Mile Island Accident," US Nuclear Regulatory Commission, June 2018, https://www.nrc.gov/reading-rm/doc-collections/fact-sheets/3mile-isle.html.

8. Ibid.

9. "Chernobyl Accident 1986," World Nuclear Organization, April 2018 (updated), http://www.world-nuclear.org/information-library/safety-and-security/safety-of-plants/chernobyl-accident.aspx.

10. Ibid.

11. Ibid.

12. *The Fukushima Daiichi Accident: Technical Volume 3/5* (Vienna: IAEA, August 2015), https://www-pub.iaea.org/MTCD/Publications/PDF/AdditionalVolumes/P1710/Pub1710-TV3-Web.pdf; Jake Adelstein, "Who's Responsible for the Fukushima Disaster?" *Japan Times*, May 2012, https://www.japantimes.co.jp/news/2015/10/03/national/media-national/whos-responsible-fukushima-disaster/#.WrzmUojwaUk.

13. Adelstein, "Who's Responsible?"

14. Nathanael Massey, "Fukushima Disaster Blame Belongs with Top Leaders at Utilities, Government and Regulators," *Scientific American*, July 6, 2012, https://www.scientificamerican.com/article/fukushima-blame-utilities-goverment-leaders-regulators/.

15. A. Hasegawa et al., "Emergency Responses and Health Consequences after the Fukushima Accident; Evacuation and Relocation," *Clinical Oncology* 28, no. 4 (April 2016): 237–44, https://www.sciencedirect.com/science/article/pii/S0936655516000054.

16. Will Kenton, "FAAMG Stocks," Investopedia, March 6, 2018, https://www.investopedia.com/terms/f/faamg-stocks.asp.

17. Alex Hern, "Facebook, Google and Twitter to Testify in Congress over Extremist Content," *Guardian*, January 10, 2018, https://www.theguardian.com/technology/2018/jan/10/facebook-google-twitter-testify-congress-extremist-content-russian-election-interference-information.

18. Taylor Hatmaker, "On Russia, Tech Doesn't Know What It Doesn't Know," TechCrunch, November 1, 2017, https://techcrunch.com/2017/11/01/russia-senate-intel-hearing-facebook-google-twitter/.

19. Ibid.

20. Ibid.

21. Stephanie J. Bird, "Socially Responsible Science Is More Than 'Good Science,'" *Journal of Microbiology & Biology Education* 15, no. 2 (December 15, 2014): 169–72, https://www.ncbi.nlm.nih.gov/pmc/articles/PMC4278471/.

22. Cortney Weinbaum, Eric Landree, Marjory Blumental, Tepring Piquado, and Carlos Ignacio Gutierrez, *Ethics in Scientific Research: An Examination of Ethical Principles and Emerging Topics* (Santa Monica, CA: RAND Corporation, forthcoming).

CHAPTER 12. REFLECTIONS ON THE FUTURE OF TECHNOLOGY

1. Sean Moffitt, "The Top 30 Emerging Technologies (2018–2028)," Medium, April 4, 2018, https://medium.com/@seanmoffitt/the-top-30-emerging-technologies-2018-2028-eca0dfb0f43c.

2. James Manyika et al., "Disruptive Technologies: Advances That Will Transform Life, Business, and the Global Economy," McKinsey Global Initiative, May 2013, https://www.mckinsey.com/business-functions/digital-mckinsey/our-insights/disruptive-technologies.

3. "Gartner Hype Cycle," Gartner, https://www.gartner.com/en/research/methodologies/gartner-hype-cycle.

APPENDIX A. THE TECHNOLOGY LEXICON

1. John Butler-Adam, "The Weighty History and Meaning behind the Word 'Science,'" *Conversation*, October 1, 2015, http://theconversation.com/the-weighty-history-and-meaning-behind-the-word-science-48280.

2. *Encyclopedia Britannica*, s.v. "Aristotle," https://www.britannica.com/biography/Aristotle.

3. *Encyclopedia Britannica*, s.v. "Natural History: Encyclopedic Scientific Work by Pliny the Elder," https://www.britannica.com/topic/Natural-History-encyclopedic-scientific-by-Pliny-the-Elder.

4. Martyn Shuttleworth, "Who Invented the Scientific Method?" Explorable, https://explorable.com/who-invented-the-scientific-method.

5. Ibid.

6. Ibid.

7. "Koch's Postulates to Identify the Causative Agent of an Infectious Disease," (PowerPoint lecture slide, University of Maryland), http://www.life.umd.edu/classroom/bsci424/BSCI223WebSiteFiles/KochsPostulates.htm.

8. "Our Definition of Science," Science Council, https://sciencecouncil.org/about-science/our-definition-of-science/.

9. *Oxford Dictionary*, s.v. "Science," https://en.oxforddictionaries.com/definition/science.

10. "Science," Vocabulary.com, https://www.vocabulary.com/dictionary/science.

11. "Science," Science Council, https://sciencecouncil.org/; Ian Sample, "What Is This Thing We Call Science? Here's One Definition . . .," *Guardian*, March 3, 2009, https://www.theguardian.com/science/blog/2009/mar/03/science-definition-council-francis-bacon.

12. "What is the Scientific Method?," Explorable, https://explorable.com/what-is-the-scientific-method.

13. L. M. Sacasas, "Traditions of Technological Criticism," *The Frailest Thing* (blog), February 15, 2014, https://thefrailestthing.com/2014/02/15/technology-that-word-you-keep-using-i-do-not-think-it-means/.

14. *Your Dictionary*, s.v. "Technology," http://www.yourdictionary.com/technology.

15. *Business Dictionary*, s.v. "Technology," http://www.businessdictionary.com/definition/technology.html.

16. *Collins Dictionary*, s.v. "Technology," https://www.collinsdictionary.com/dictionary/english/technology.

17. The definitions for the technology readiness levels come from "Technology Readiness Levels in the Department of Defense," in *Technologies to Enable Autonomous Detection for BioWatch*, eds. India Hook-Barnard, Sheena M. Posey Norris, and Joe Alper (Washington, DC: National Academies Press, April 2011), https://www.ncbi.nlm.nih.gov/books/NBK201356/.

18. From GAO-12-837 report on Department of Homeland Security (DHS) Research and Development, the definitions are: (1) Basic research is a systematic study directed toward a fuller knowledge or understanding of the fundamental aspects of phenomena and of observable facts without specific applications toward processes or products in mind. (2) Applied research is a systematic study to gain knowledge or understanding to determine the means by which a recognized and specific need may be met. (3) Development is a systematic application of knowledge or understanding, directed toward the production of useful materials, devices, and systems or methods, including design, development, and improvement of prototypes and new processes to meet specific requirements.

19. "Technology Readiness Levels," National Institutes of Health, US Department of Health and Human Services, updated August 2016, https://ncai.nhlbi.nih.gov/ncai/resources/techreadylevels.

20. *Business Dictionary,* s.v. "Innovation," http://www.businessdictionary.com/definition/innovation.html.

21. Ibid.

22. Fred Kaplan, "The Pentagon's Innovation Experiment," *MIT Technology Review*, December 19, 2016, https://www.technologyreview.com/s/603084/the-pentagons-innovation-experiment/.

23. "Formulation of Organizational Transformation Strategies," MITRE, https://www.mitre.org/publications/systems-engineering-guide/enterprise-engineering/transformation-planning-and-organizational-change/formulation-of-organizational-transformation-strategies.

24. These examples all came from Josh Sanburn, "10 Companies That Radically Transformed Their Businesses," *Time*, October 19, 2011, http://business.time.com/2011/06/16/10-companies-that-radically-transformed-their-businesses/.

25. "National Security Act," National Security Archives, https://catalog.archives.gov/id/299856.

26. *Joint Vision 2010* (Washington, DC: Office of the Chairman of the Joint Chiefs of Staff, 1995), http://webapp1.dlib.indiana.edu/virtual_disk_library/index.cgi/4240529/FID378/pdfdocs/2010/Jv2010.pdf.

APPENDIX B. TRL DEFINITIONS, DESCRIPTIONS, AND SUPPORTING INFORMATION FROM THE US DEPARTMENT OF DEFENSE

1. "Technology Readiness Levels in the Department of Defense," in *Technologies to Enable Autonomous Detection for BioWatch*, eds. India Hook-Barnard, Sheena M. Posey Norris, and Joe Alper (Washington, DC: National Academies Press, April 2011), https://www.ncbi .nlm.nih.gov/books/NBK201356/.

INDEX

ABOUT THE AUTHOR

Dr. Daniel Gerstein is a national security professional and technology expert who has served the United States not only in senior government positions but also in uniform, in industry, in academia, and in think tanks. As a senior government civilian official, he was acting undersecretary and deputy undersecretary in the Department of Homeland Security's Science and Technology Directorate. In this position, he directed the organization's annual budget of more than $1 billion. During his time in the US Army, he served on four continents while participating in combat, peacekeeping, humanitarian assistance, and homeland security efforts. He also held high-level assignments in the Pentagon for more than a decade. Dr. Gerstein has extensive experience with international negotiations, notably as a member of the Holbrooke Delegation that negotiated the peace settlement in Bosnia. He is a frequent news contributor, and he has published numerous books, articles, and commentaries on a wide variety of national and homeland security issues. He also is a member of several corporate boards and advisory committees.